Multi-Robot Systems. From Swarms to Intelligent Automata
Volume III

Multi-Robot Systems. From Swarms to Intelligent Automata Volume III

Proceedings from the 2005 International Workshop on Multi-Robot Systems

Edited by

LYNNE E. PARKER
The University of Tennessee,
Knoxville, TN, U.S.A.

FRANK E. SCHNEIDER
FGAN, Wachtberg, Germany

and

ALAN C. SCHULTZ
Navy Center for Applied Research in A.I.,
Naval Research Laboratory,
Washington, DC, U.S.A.

Springer

A C.I.P. Catalogue record for this book is available from the Library of Congress.

ISBN-10 1-4020-3388-5 (HB) Springer Dordrecht, Berlin, Heidelberg, New York
ISBN-10 1-4020-3389-3 (e-book) Springer Dordrecht, Berlin, Heidelberg, New York
ISBN-13 978-1-4020-3388-9 (HB) Springer Dordrecht, Berlin, Heidelberg, New York
ISBN-13 978-1-4020-3389-6 (e-book) Springer Dordrecht, Berlin, Heidelberg, New York

---

Published by Springer,
P.O. Box 17, 3300 AA Dordrecht, The Netherlands.

Printed on acid-free paper

Printed in the Netherlands.

Contents

Preface

The Third International Workshop on Multi-Robot Systems was held in March 2005 at the Naval Research Laboratory in Washington, D.C., USA. Bringing together leading researchers and government sponsors for three days of technical interchange on multi-robot systems, the workshop follows two previous highly successful gatherings in 2002 and 2003.Like the previous two workshops, the meeting began with presentations by various government program managers describing application areas and programs with an interest in multi-robot systems. U.S. Government representatives were on hand from the Office of Naval Research and several other governmental offices.Top researchers in the field then presented their current activities in many areas of multi-robot systems. Presentations spanned a wide range of topics, including task allocation, coordination in dynamic environments, information/sensor sharing and fusion, distributed mapping and coverage, motion planning and control, human-robot interaction, and applications of multi-robot systems. All presentations were given in a single-track workshop format. This proceedings documents the work presented at the workshop.The research presentations were followed by panel discussions, in which all participants interacted to highlight the challenges of this field and to develop possible solutions. In addition to the invited research talks, researchers and students were given an opportunity to present their work at poster sessions.We would like to thank the Naval Research Laboratory for sponsoring this workshop and providing the facilities for these meetings to take place.We are extremely grateful to Magdalena Bugajska, Paul Wiegand, and Mitchell A. Potter, for their vital help (and long hours) in editing these proceedings and to Michelle Caccivio for providing the administrative support to the workshop.

LYNNE E. PARKER, ALAN C. SCHULTZ, AND FRANK E. SCHNEIDER

I

TASK ALLOCATION

THE GENERATION OF BIDDING RULES FOR AUCTION-BASED ROBOT COORDINATION *

Craig Tovey, Michail G. Lagoudakis
School of Industrial and Systems Engineering, Georgia Institute of Technology
{ctovey, michail.lagoudakis} @isye.gatech.edu

Sonal Jain, Sven Koenig
Computer Science Department, University of Southern California
{sonaljai, skoenig} @usc.edu

Abstract Robotics researchers have used auction-based coordination systems for robot teams because of their robustness and efficiency. However, there is no research into systematic methods for deriving appropriate bidding rules for given team objectives. In this paper, we propose the first such method and demonstrate it by deriving bidding rules for three possible team objectives of a multi-robot exploration task. We demonstrate experimentally that the resulting bidding rules indeed exhibit good performance for their respective team objectives and compare favorably to the optimal performance. Our research thus allows the designers of auction-based coordination systems to focus on developing appropriate team objectives, for which good bidding rules can then be derived automatically.

Keywords: Auctions, Bidding Rules, Multi-Robot Coordination, Exploration.

1. Introduction

The time required to reach other planets makes planetary surface exploration missions prime targets for automation. Sending rovers to other planets either instead of or together with people can also significantly reduce the danger and cost involved. Teams of rovers are both more fault tolerant (through redundancy) and more efficient (through parallelism) than single rovers if the rovers are coordinated well. However, rovers cannot be easily tele-operated since this

*We thank Apurva Mudgal for his help. This research was partly supported by NSF awards under contracts ITR/AP0113881, IIS-0098807, and IIS-0350584. The views and conclusions contained in this document are those of the authors and should not be interpreted as representing the official policies, either expressed or implied, of the sponsoring organizations, agencies, companies or the U.S. government.

L.E. Parker et al. (eds.),
Multi-Robot Systems. From Swarms to Intelligent Automata. Volume III, 3–14.

requires a large number of human operators and is communication intensive, error prone, and slow. Neither can they be fully preprogrammed since their activities depend on their discoveries. Thus, one needs to endow them with the capability to coordinate autonomously with each other. Consider, for example, a multi-robot exploration task where a team of lunar rovers has to visit a number of given target locations to collect rock samples. Each target must be visited by at least one rover. The rovers first allocate the targets to themselves, and each rover then visits the targets that are allocated to it. The rovers know their current location at all times but might initially not know where obstacles are in the terrain. It can therefore be beneficial for the rovers to re-allocate the targets to themselves as they discover more about the terrain during execution, for example, when a rover discovers that it is separated by a big crater from its next target. Similar multi-robot exploration tasks arise for mine sweeping, search and rescue operations, police operations, and hazardous material cleaning, among others.

Multi-robot coordination tasks are typically solved with heuristic methods since optimizing the performance is often computationally intractable. They are often solved with decentralized methods since centralized methods lack robustness: if the central controller fails, so does the entire robot team. Market mechanisms, such as auctions, are popular decentralized and heuristic multi-robot coordination methods (Rabideau et al., 2000). In this case, the robots are the bidders and the targets are the goods up for auction. Every robot bids on targets and then visits all targets that it wins. As the robots discover more about the terrain during execution, they run additional auctions to change the allocation of targets to themselves. The resulting auction-based coordination system is efficient in terms of communication (robots communicate only numeric bids) and computation (robots compute their bids in parallel). It is therefore not surprising that auctions have been shown to be effective multi-robot coordination methods (Gerkey and Matarić, 2002, Zlot et al., 2002, Thayer et al., 2000, Goldberg et al., 2003). However, there are currently no systematic methods for deriving appropriate bidding rules for given team objectives. In this paper, we propose the first such method and demonstrate it by deriving bidding rules for three possible team objectives of the multi-robot exploration task. We demonstrate experimentally that the resulting bidding rules indeed exhibit good performance for their respective team objectives and compare favorably to the optimal performance. Our research thus allows the designers of auction-based coordination systems to focus on developing appropriate team objectives, for which good bidding rules can then be derived automatically.

2. The Auction-Based Coordination System

In known environments, all targets are initially unallocated. During each round of bidding, all robots bids on all unallocated targets. The robot that places the overall lowest bid on any target is allocated that particular target. A new round of bidding starts, and all robots bid again on all unallocated targets, and so on until all targets have been allocated to robots. (Note that each robot needs to bid only on a single target during each round, namely on one of the targets for which its bid is the lowest, since all other bids from the same robot have no chance of winning.) Each robot then calculates the optimal path for the given team objective for visiting the targets allocated to it and then moves along that path. A robot does not move if no targets are allocated to it.

In unknown environments, the robots proceed in the same way but under the optimistic initial assumption that there are no obstacles. As the robots move along their paths and a robot discovers a new obstacle, it informs the other robots about it. Each robot then re-calculates the optimal path for the given team objective for visiting the unvisited targets allocated to it, taking into account all obstacles that it knows about. If the performance significantly degrades for at least one robot (in our experiments, we use a threshold of 10 percent difference), then the robots use auctions to re-allocate all unvisited targets among themselves. Each robot then calculates the optimal path for the given team objective for visiting the targets allocated to it and then moves along that path, and so on until all targets have been visited.

This auction-based coordination system is similar to multi-round auctions and sequential single-item auctions. Its main advantage is its simplicity and the fact that it allows for a decentralized implementation on real robots. Each robot computes its one bid locally and in parallel with the other robots, broadcasts the bid to the other robots, listens to the broadcasts of the other robots, and then locally determines the winning bid. Thus, there is no need for a central auctioneer and therefore no single point of failure. A similar but more restricted auction scheme has been used in the past for robot coordination (Dias and Stentz, 2000).

3. Team Objectives for Multi-Robot Exploration

A multi-robot exploration task consists of the locations of n robots and m targets as well as a cost function that specifies the cost of moving between locations. The objective of the multi-robot exploration task is to find an allocation of targets to robots and a path for each robot that visits all targets allocated to it so that the team objective is achieved. Note that the robots are not required to return to their initial locations. In this paper, we study three team objectives:

- MINISUM: Minimize the sum of the path costs over all robots.

- MINIMAX: Minimize the maximum path cost over all robots.
- MINIAVE: Minimize the average per target cost over all targets.

The path cost of a robot is the sum of the costs along its path, from its initial location to the first target on the path, and so on, stopping at the last target on the path. The per target cost of a target is the sum of the costs along the path of the robot that visits the target in question, from its initial location to the first target on the path, and so on, stopping at the target in question.

Optimizing the performance for the three team objectives is NP-hard and thus likely computationally intractable, as they resemble the Traveling Salesperson Problem, the Min-Max Vehicle Routing Problem, and the Traveling Repairperson Problem (or Minimum Latency Problem), respectively, which are intractable even on the Euclidean plane. However, these team objectives cover a wide range of applications. For example, if the cost is energy consumption, then the MINISUM team objective minimizes the total energy consumed by all robots until all targets have been visited. If the cost is travel time, then the MINIMAX team objective minimizes the time until all targets have been visited (task-completion time) and the MINIAVE team objective minimizes how long it takes on average until a target is visited (target-visit time). The MINISUM and MINIMAX team objectives have been used in the context of multi-robot exploration (Dias and Stentz, 2000, Dias and Stentz, 2002, Berhault et al., 2003, Lagoudakis et al., 2004). The MINIAVE team objective, on the other hand, has not been used before in this context although it is very appropriate for search-and-rescue tasks, where the health condition of several victims deteriorates until a robot visits them. Consider, for example, an earthquake scenario where an accident site with one victim is located at a travel time of 20 units to the west of a robot and another accident site with twenty victims is located at a travel time of 25 units to its east. In this case, visiting the site to the west first and then the site to the east achieves both the MINISUM and the MINIMAX team objectives. However, the twenty victims to the east are visited very late and their health condition thus is very bad. On the other hand, visiting the site to the east first and then the site to the west achieves the MINIAVE team objective and results in an overall better average health condition of the victims. This example illustrates the importance of the MINIAVE team objective in cases where the targets occur in clusters of different sizes.

4. Systematic Generation of Bidding Rules

We seek to derive an appropriate bidding rule for a given team objective. This problem has not been studied before in the robotics literature. Assume that there are n robots $r_1, \ldots, r_n$ and m currently unallocated targets $t_1, \ldots, t_m$. Assume further that the team objective has the structure to assign a set of targets T_i to robot r_i for all i, where the sets $T = \{T_1, \ldots, T_n\}$ form a partition of all targets that optimizes the performance $f\big(g(r_1,T_1),\ldots,g(r_n,T_n)\big)$ for given functions f and g. Function g determines the performance of each robot, and function f determines the performance of the team as a function of the performance of the robots. The three team objectives fit this structure. For any robot r_i and any set of targets T_i, let $PC(r_i,T_i)$ denote the minimum path cost of robot r_i and $STC(r_i,T_i)$ denote the minimum sum of per target costs over all targets in T_i if robot r_i visits all targets in T_i from its current location. Then, it holds that

- MINISUM: $\min_T \sum_j PC(r_j,T_j)$,
- MINIMAX: $\min_T \max_j PC(r_j,T_j)$, and
- MINIAVE: $\min_T \frac{1}{m}\sum_j STC(r_j,T_j)$.

A bidding rule determines how much a robot bids on a target. We propose the following bidding rule for a given team objective, which is directly derived from the team objective itself.

Bidding Rule Robot r bids on target t the difference in performance for the given team objective between the current allocation of targets to robots and the allocation that results from the current one if robot r is allocated target t. (Unallocated targets are ignored.)

Consequently, robot r_i should bid on target t

$$f\big(g(r_1,T_1'),\ldots,g(r_n,T_n')\big)-f\big(g(r_1,T_1),\ldots,g(r_n,T_n)\big),$$

where $T_i' = T_i \cup \{t\}$ and $T_j' = T_j$ for $i \neq j$. The bidding rule thus performs hill climbing to maximize the performance and can thus suffer from local optima. However, optimizing the performance is NP-hard for the three team objectives. Our auction-based coordination system is therefore not designed to optimize the performance but to be efficient and result in a good performance, and hill climbing has these properties. One potential problem with the bidding rule is that the robots might not have all the information needed to compute the bids. For example, a robot may not know the locations of the other robots. However, we will now show that a robot can calculate its bids for the three team objectives knowing only its current location, the set of targets allocated to it, and the cost function:

- For the MINISUM team objective, robot r_i should bid on target t

$$\sum_j PC(r_j,T_j') - \sum_j PC(r_j,T_j) = PC(r_i,T_i\cup\{t\}) - PC(r_i,T_i).$$

- For the MINIMAX team objective, robot r_i should bid on target t

$$\max_j PC(r_j, T'_j) - \max_j PC(r_j, T_j) = PC(r_i, T_i \cup \{t\}) - \max_j PC(r_j, T_j).$$

This derivation uses the fact that $\max_j PC(r_j, T'_j) = PC(r_i, T'_i)$, otherwise target t would have already been allocated in a previous round of bidding. The term $\max_j PC(r_j, T_j)$ can be dropped since the outcomes of the auctions remain unchanged if all bids change by a constant. Thus, robot r_i can bid just $PC(r_i, T_i \cup \{t\})$ on target t.

- For the MINIAVE team objective, robot r_i should bid on target t

$$\frac{1}{m}\sum_j STC(r_j, T'_j) - \frac{1}{m}\sum_j STC(r_j, T_j) = \frac{1}{m}\Big(STC(r_i, T_i \cup \{t\}) - STC(r_i, T_i)\Big).$$

The factor $1/m$ can be dropped since the outcomes of the auctions remain unchanged if all bids are multiplied by a constant factor. Thus, robot r_i can bid just $STC(r_i, T_i \cup \{t\}) - STC(r_i, T_i)$ on target t.

Thus, the bidding rules for the three team objectives are

- BIDSUM: $PC(r_i, T_i \cup \{t\}) - PC(r_i, T_i)$,
- BIDMAX: $PC(r_i, T_i \cup \{t\})$, and
- BIDAVE: $STC(r_i, T_i \cup \{t\}) - STC(r_i, T_i)$.

The robots need to be able to calculate their bids efficiently but computing $PC(r_i, T_i \cup \{t\})$ or $STC(r_i, T_i \cup \{t\})$ is NP-hard. Robot r_i thus uses a greedy method to approximate these values. In particular, it finds a good path that visits the targets in $T_i \cup \{t\}$ for a given team objective as follows. It already has a good path that visits the targets in T_i. First, it inserts target t into all positions on the existing path, one after the other. Then, it tries to improve each new path by first using the 2-opt improvement rule and then the 1-target 3-opt improvement rule. Finally, it picks the best one of the resulting paths for the given team objective. The 2-opt improvement rule takes a path and inverts the order of targets in each one of its continuous subpaths in turn, picks the best one of the resulting paths for the given team objective, and repeats the procedure until the path can no longer be improved. The 1-target 3-opt improvement rule removes a target from the path and inserts it into all other possible positions on the path, picks the best one of the resulting paths for the given team objective, and repeats the procedure until the path can no longer be improved.

The three bidding rules are not guaranteed to achieve their respective team objectives even if the values $PC(r_i, T_i \cup \{t\})$ and $STC(r_i, T_i \cup \{t\})$ are computed exactly. Consider the simple multi-robot exploration task in Figure 1 with 2 robots and 2 targets and unit costs between adjacent locations. All bidding rules

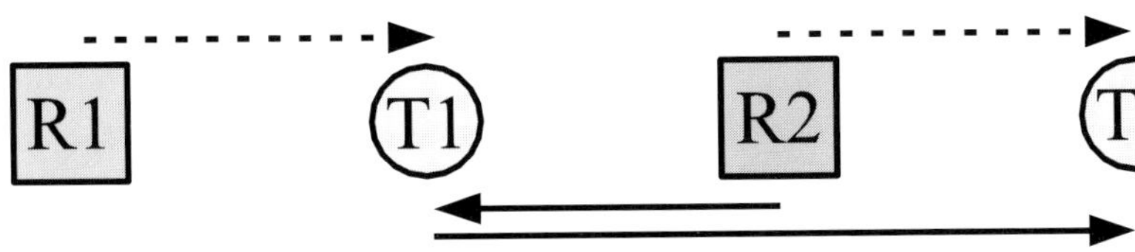

Figure 1. A simple multi-robot exploration task.

can result in the robots following the solid lines, resulting in a performance of 3 for the MINISUM team objective, a performance of 3 for the MINIMAX team objective, and a performance of 2 for the MINIAVE team objective. However, the robots should follow the dashed lines to maximize the performance for all three team objectives, resulting in a performance of 2 for the MINISUM team objective, a performance of 1 for the MINIMAX team objective, and a performance of 1 for the MINIAVE team objective. (We rely on a particular way of breaking ties in this multi-robot exploration example but can easily change the edge costs by small amounts to guarantee that the bidding rules result in the robots following the solid lines independently of how ties are broken.) In a forthcoming paper, we analyze the performance of the three bidding rules theoretically and show that the performance of the BIDSUM bidding rule in the Euclidean case is at most a factor of two away from optimum, whereas no constant-factor bound exists for the performance of the BIDMAX and BIDAVE bidding rules even in the Euclidean case.

5. Experimental Evaluation

To demonstrate that the performance of the three bidding rules is indeed good for their respective team objectives, we implemented them and then tested them in office-like environments with rooms, doors, and corridors, as shown in Figure 2. We performed experiments with both unclustered and clustered targets. The locations of the robots and targets for each multi-robot exploration task were chosen randomly in the unclustered target case. The locations of the robots and targets were also chosen randomly in the clustered target case, but with the restriction that 50 percent of the targets were placed in clusters of 5 targets each. The numbers in the tables below are averages over 10 different multi-robot exploration tasks with the same settings. The performance of the best bidding rule for a given team objective is shown in bold.

5.1 Known Environments

We mapped our environments onto eight-connected uniform grids of size 51×51 and computed all costs between locations as the shortest distances on the grid. Our auction-based coordination system used these costs to find an allocation of targets to robots and a path for each robot that visits all targets allocated to it. We interfaced it to the popular Player/Stage robot simulator (Gerkey et al., 2003) to execute the paths and visualize the resulting robot trails. Figure 2 shows the initial locations of the robots (squares) and targets (circles) as well as the resulting robot trails (dots) for each one of the three bidding rules for a sample multi-robot exploration task with 3 robots and 20 unclustered targets in a completely known environment. SUM, MAX and AVE in the caption of the figure denote the performance for the MINISUM, MINIMAX and MINIAVE team objectives, respectively. Each bidding rule results in a better performance for its team objective than the other two bidding rules. For example, the BIDSUM bidding rule results in paths of very different lengths, whereas the BIDMAX bidding rule results in paths of similar lengths. Therefore, the performance of the BIDMAX bidding rule is better for the MINIMAX team objective than the one of the BIDSUM bidding rule.

We compared the performance of the three bidding rules against the optimal performance for multi-robot exploration tasks with one or two robots and ten targets. The optimal performance was calculated by formulating the multi-robot exploration tasks as integer programs and solving them with the commercial mixed integer program solver CPLEX. The NP-hardness of optimizing the performance did not allow us to solve larger multi-robot exploration tasks. Table 1 shows the performance of each bidding rule and the optimal performance for each team objective. Again, each bidding rule results in a better performance for its team objective than the other two bidding rules, with the exception of ties between the BIDSUM and BIDMAX bidding rules for multi-robot exploration tasks with one robot. These ties are unavoidable because the MINISUM and MINIMAX team objectives are identical for one-robot exploration tasks. The performance of the best bidding rule for each team objective is always close to the optimal performance. In particular, the performance of the BIDSUM bidding rule for the MINISUM team objective is within a factor of 1.10 of optimal, the performance of the BIDMAX bidding rule for the MINIMAX team objective is within a factor of 1.44 of optimal, and the performance of the BIDAVE bidding rule for the MINIAVE team objective is within a factor of 1.28 of optimal.

We also compared the performance of the three bidding rules against each other for large multi-robot exploration tasks with one, five or ten robots and 100 targets. Table 2 shows the performance of each bidding rule. Again, each bidding rule results in a better performance for its team objective than the other two bidding rules, with the exception of the unavoidable ties.

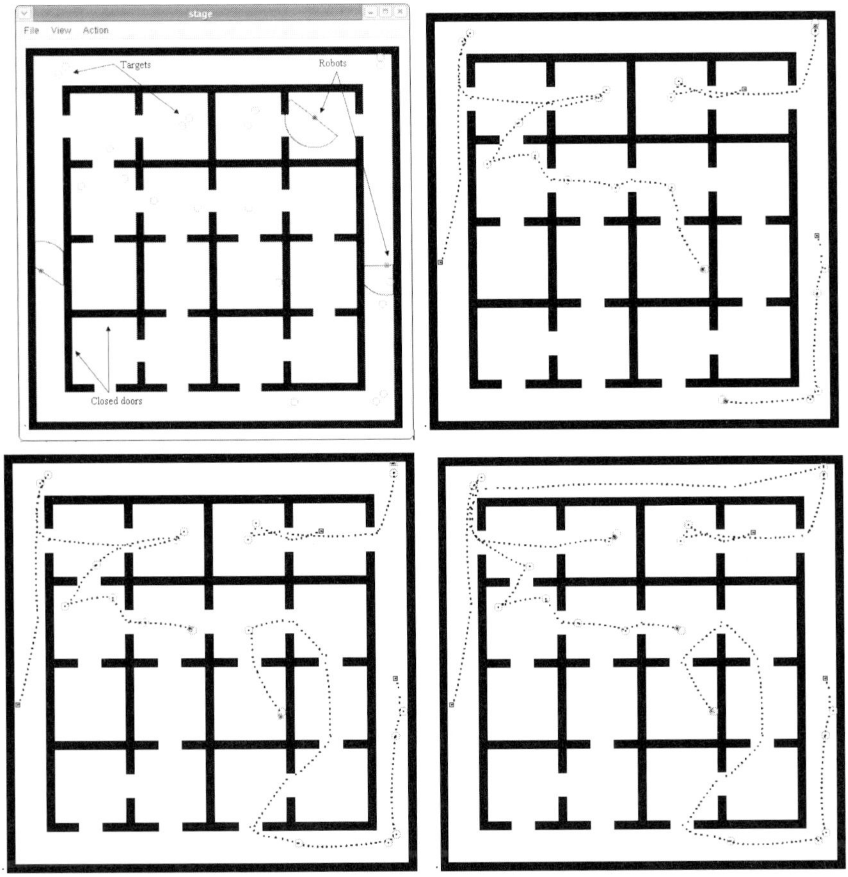

Figure 2. Player/Stage screenshots: initial locations (top left) and robot trails with the BIDSUM (top right) [SUM=182.50, MAX=113.36 , AVE=48.61], BIDMAX (bottom left) [SUM=218.12 , MAX=93.87 , AVE=46.01], and BIDAVE (bottom right) [SUM=269.27 , MAX=109.39 , AVE=45.15] bidding rules.

5.2 Unknown Environments

We compared the performance of the three bidding rules against each other for the same large multi-robot exploration tasks as in the previous section but in initially completely unknown environments. In this case, we mapped our environments onto four-connected uniform grids of size 51×51 and computed all costs between locations as the shortest distances on the grid. These grids were also used to simulate the movement of the robots in a coarse and noise-free simulation. (We could not use eight-connected grids because diagonal movements are longer than horizontal and vertical ones, and the simulation steps thus would need to be much smaller than moving from cell to cell.) The robots sense all blockages in their immediate four-cell neighborhood. Table 3 shows

Table 1. Performance of bidding rules against optimal in known environments.

Robots	Bidding Rule	Unclustered SUM	Unclustered MAX	Unclustered AVE	Clustered SUM	Clustered MAX	Clustered AVE
1	BIDSUM	**199.95**	**199.95**	103.08	**143.69**	**143.69**	78.65
1	BIDMAX	**199.95**	**199.95**	103.08	**143.69**	**143.69**	78.65
1	BIDAVE	214.93	214.93	**98.66**	155.50	155.50	**63.12**
1	OPTIMAL	199.95	199.95	98.37	143.69	143.69	63.12
2	BIDSUM	**193.50**	168.50	79.21	**134.18**	97.17	62.47
2	BIDMAX	219.15	**125.84**	61.39	144.84	**90.10**	57.38
2	BIDAVE	219.16	128.45	**59.12**	157.29	100.56	**49.15**
2	OPTIMAL	189.15	109.34	55.45	132.06	85.86	47.63

Table 2. Performance of bidding rules against each other in known environments.

Robots	Bidding Rule	Unclustered SUM	Unclustered MAX	Unclustered AVE	Clustered SUM	Clustered MAX	Clustered AVE
1	BIDSUM	**554.40**	**554.40**	281.11	**437.25**	**437.25**	212.81
1	BIDMAX	**554.40**	**554.40**	281.11	**437.25**	**437.25**	212.81
1	BIDAVE	611.50	611.50	**243.30**	532.46	532.46	**169.20**
5	BIDSUM	**483.89**	210.30	80.74	**374.33**	186.50	66.94
5	BIDMAX	548.40	**130.41**	58.70	450.72	**112.18**	50.50
5	BIDAVE	601.28	146.18	**55.19**	500.05	132.98	**42.41**
10	BIDSUM	**435.30**	136.70	45.89	**318.52**	102.15	35.14
10	BIDMAX	536.90	**77.95**	31.39	402.30	**63.89**	25.88
10	BIDAVE	564.73	88.23	**30.04**	437.23	71.52	**22.02**

the performance of each bidding rule. Again, each bidding rule results in a better performance for its team objective than the other two bidding rules, with the exception of the unavoidable ties and two other exceptions. The average number of auctions is 28.37 with a maximum of 82 auctions in one case. In general, the number of auctions increases with the number of robots. Note that the difference in performance between known and unknown environments is at most a factor of three. It is remarkable that our auction-based coordination system manages to achieve such a good performance for all team objectives since there has to be some performance degradation given that we switched both from known to unknown environments and from eight-connected to four-connected grids.

6. Conclusions and Future Work

In this paper, we described an auction-based coordination system and then proposed a systematic method for deriving appropriate bidding rules for given

Table 3. Performance of bidding rules against each other in unknown environments.

Robots	Bidding Rule	Unclustered SUM	Unclustered MAX	Unclustered AVE	Clustered SUM	Clustered MAX	Clustered AVE
1	BIDSUM	**1459.90**	**1459.90**	**813.40**	**1139.20**	**1139.20**	672.14
1	BIDMAX	**1459.90**	**1459.90**	**813.40**	**1139.20**	**1139.20**	672.14
1	BIDAVE	1588.50	1588.50	826.82	1164.40	1164.40	**463.14**
5	BIDSUM	**943.60**	586.90	223.47	**771.40**	432.90	166.60
5	BIDMAX	979.00	**238.10**	98.48	811.30	216.90	86.58
5	BIDAVE	992.10	240.10	**90.54**	838.30	**214.10**	**79.36**
10	BIDSUM	**799.50**	312.20	93.69	**596.10**	223.20	63.95
10	BIDMAX	885.40	**123.60**	48.43	677.80	**110.60**	37.92
10	BIDAVE	871.80	133.00	**45.19**	697.80	121.50	**35.43**

team objectives. We then demonstrated it by deriving bidding rules for three possible team objectives of a multi-robot exploration task, that relate to minimizing the total energy consumption, task-completion time, and average target-visit time. (The last team objective had not been used before but we showed it to be appropriate for search-and-rescue tasks.) Finally, we demonstrated experimentally that the derived bidding rules indeed exhibit good performance for their respective team objectives and compare favorably to the optimal performance. In the future, we intend to adapt our methodology to other multi-robot coordination tasks. For example, we intend to study multi-robot coordination with auction-based coordination systems in the presence of additional constraints, such as compatibility constraints which dictate that certain targets can only be visited by certain robots.

References

Berhault, M., Huang, H., Keskinocak, P., Koenig, S., Elmaghraby, W., Griffin, P., and Kleywegt, A. (2003). Robot exploration with combinatorial auctions. In *Proceedings of the International Conference on Intelligent Robots and Systems*, pages 1957–1962.

Dias, M. and Stentz, A. (2000). A free market architecture for distributed control of a multirobot system. In *Proceedings of the International Conference on Intelligent Autonomous Systems*, pages 115–122.

Dias, M. and Stentz, A. (2002). Enhanced negotiation and opportunistic optimization for market-based multirobot coordination. Technical Report CMU-RI-TR-02-18, Robotics Institute, Carnegie Mellon University, Pittsburgh (Pennsylvania).

Gerkey, B. and Matarić, M. (2002). Sold!: Auction methods for multi-robot coordination. *IEEE Transactions on Robotics and Automation*, 18(5):758–768.

Gerkey, B., Vaughan, R., Stoy, K., Howard, A., Sukhatme, G., and Matarić, M. (2003). Most valuable player: A robot device server for distributed control. In *Proceedings of the International Conference on Intelligent Robots and Systems*, pages 1226–1231.

Goldberg, D., Circirello, V., Dias, M., Simmons, R., Smith, S., and Stentz, A. (2003). Market-based multi-robot planning in a distributed layered architecture. In *Proceedings from the International Workshop on Multi-Robot Systems*, pages 27–38.

Lagoudakis, M., Berhault, M., Keskinocak, P., Koenig, S., and Kleywegt, A. (2004). Simple auctions with performance guarantees for multi-robot task allocation. In *Proceedings of the International Conference on Intelligent Robots and Systems.*

Rabideau, G., Estlin, T., Chien, S., and Barrett, A. (2000). A comparison of coordinated planning methods for cooperating rovers. In *Proceedings of the International Conference on Autonomous Agents*, pages 100–101.

Thayer, S., Digney, B., Dias, M., Stentz, A., Nabbe, B., and Hebert, M. (2000). Distributed robotic mapping of extreme environments. In *Proceedings of SPIE: Mobile Robots XV and Telemanipulator and Telepresence Technologies VII*, volume 4195, pages 84–95.

Zlot, R., Stentz, A., Dias, M., and Thayer, S. (2002). Multi-robot exploration controlled by a market economy. In *Proceedings of the International Conference on Robotics and Automation*, pages 3016–3023.

ISSUES IN MULTI-ROBOT COALITION FORMATION

Lovekesh Vig
Electrical Engineering and Computer Science Department
Vanderbilt University, Nashville TN 37212
lovekesh.vig@vanderbilt.edu

Julie A. Adams
Electrical Engineering and Computer Science Department
Vanderbilt University, Nashville TN 37212
julie.a.adams@vanderbilt.edu

Abstract Numerous coalition formation algorithms exist in the Distributed Artificial Intelligence literature. Algorithms exist that form agent coalitions in both super additive and non-super additive environments. The employed techniques vary from negotiation-based protocols in Multi-Agent System (MAS) environments to those based on computation in Distributed Problem Solving (DPS) environments. Coalition formation behaviors have also been discussed in the game theory literature.

Despite the plethora of multi-agent coalition formation literature, to the best of our knowledge none of these algorithms have been demonstrated with an actual multiple-robot system. There exists a discrepancy between the multi-agent algorithms and their applicability to the multiple-robot domain. This work aims to correct that discrepancy by unearthing issues that arise while attempting to tailor these algorithms to the multiple-robot domain. A well-known multiple-agent coalition formation algorithm has been studied in order to identify the necessary modifications to facilitate its application to the multiple-robot domain.

Keywords: Coalition formation, fault-tolerance, multi-robot, task allocation.

1. Introduction

Multi-agent systems often encounter situations that require agents to cooperate and perform a task. In such situations it is often beneficial to assign a group of agents to a task, such as when a single agent cannot perform the tasks. This paper investigates allocating tasks to disjoint robot teams, referred to as

L.E. Parker et al. (eds.),
Multi-Robot Systems. From Swarms to Intelligent Automata. Volume III, 15–26.

coalitions. Choosing the optimal coalition from all possible coalitions is an intractable problem due to the size of coalition structure space (Sandholm et al., 1999). Algorithms exist that yield solutions within a bound from the optimal and are tractable. However these algorithms make underlying assumptions that are not applicable to the multiple-robot domain, hence the existence of a discrepancy between the multi-agent and multiple-robot coalition formation literature. This paper identifies these assumptions and provides modifications to the multi-agent coalition formation algorithms to facilitate their application in the multiple-robot domain. Gerkey and Mataric (Gerkey and Mataric, 2004) indicate that despite the existence of various multi-agent coalition formation algorithms, none of these algorithms have been demonstrated in the multiple-robot domain.

Various task allocation schemes exist. The ALLIANCE (Parker, 1998) architecture uses motivational behaviors to monitor task progress and dynamically reallocate tasks. The MURDOCH (Gerkey and Mataric, 2002) and BLE (Werger and Mataric, 2000) systems use a Publish/ Subscribe method to allocate tasks that are hierarchically distributed. However, most current task allocation schemes assume that all of the system robots are available for task execution. These systems also assume that communication between robots is always possible or that the system can provide motivational feedback. These assumptions need not always hold, a set of tasks may be located at considerable distances from one another so that the best solution is to dispatch a robot team to each designated task area and hope that the team can autonomously complete the task. The robots must then coalesce into teams responsible for each task. The focus of this work is to investigate the various issues that arise while attempting to form multiple-robot coalitions using existing multi-agent coalition formation algorithms. Some solutions are suggested and Shehory and Krauss' (Shehory and Krauss, 1998) multi-agent task allocation scheme algorithm is modified to operate in the multiple-robot domain. This algorithm was chosen because it is designed for DPS Environments, has an excellent real-time response and has been shown to provide results within a bound from optimal.

This paper is organized as follows. Section 2 provides the related work. Section 3 presents an overview of Shehory and Krauss' algorithm. Section 4 identifies issues that entail modification of current coalition formation algorithms. Experimental results are provided in Section 5. Finally, Section 6 discusses the conclusions and future work.

2. Related Work

Shehory and Krauss proposed a variety of algorithms for agent coalition formation that efficiently yield solutions close to optimal. They describe a Kernel oriented model for coalition formation in general environments (Shehory and Krauss, 1996) and non-super additive environments (Shehory and Krauss, 1999). They also provided a computation based algorithm for non-super additive environments (Shehory and Krauss, 1998). Brooks and Durfee (Brooks and Durfee, 2003) provide a novel algorithm in which selfish agents learn to form congregations. Anderson et al. (Anderson et al., 2004) discuss the formation of dynamic coalitions in robotic soccer environments by agents that can learn each other's capabilities. Fass (Fass, 2004) provides results for an Automata-theoretic view of agent coalitions that can adapt to selecting groups of agents. Li and Soh (Li and Soh, 2004) discuss the use of a reinforcement learning approach where agents learn to form better coalitions. Sorbella et al. (Sorbella et al., 2004) describe a mechanism for coalition formation based on a political society.

3. Shehory and Krauss' Algorithm

Shehory and Krauss (Shehory and Krauss, 1998) developed a multi-agent algorithm that is designed for task allocation via agent coalition formation in DPS environments.

3.1 Assumptions

The algorithm includes various assumptions. Assume a set of n agents, $N = A_1, A_2 ..., A_n$. The agents communicate with each other and are aware of all tasks to be performed. Each agent has a vector of real non-negative capabilities $B_i =< b^i_1, b^i_2 ..., b^i_r >$. Each capability quantifies the ability to perform an action. In order to assess coalitions and task execution, an evaluation function is attached to each capability type that transforms capability units into monetary units. It is assumed that there is a set of m independent tasks $T = t_1, t_2 ..., t_m$. A capability vector $B_l =< b^l_1, ..., b^l_r >$ is necessary for the satisfaction of each task t_l. The utility gained from performing the task depends on the capabilities required for its execution. A coalition is a group of agents that decide to cooperate in order to achieve a common task. Each coalition works on a single task. A coalition C has a capability vector B_c representing the sum of the capabilities that the coalition members contribute to this specific coalition. A coalition C can perform a task t only if the capability vector necessary for task fulfillment B_t satisfies $\forall 0 \leq i \leq r, b^t_i < b^c_i$.

3.2 The algorithm

The algorithm consists of two primary stages. The first calculates coalitional values to enable comparison of coalitions. The second stage entails an iterative greedy process through which the agents determine the preferred coalitions and form them. Stage one is the more relevant to this work. During this stage the evaluation of coalitions is distributed amongst the agents via extensive message passing, requiring considerable communication between agents. After this stage, each agent has a list of coalitions for which it calculated coalition values. Each agent also has all necessary information regarding the coalition memberships' capabilities. In order to calculate the coalition values, each agent then:

1 Determines the eligible coalitions for each task execution t_i by comparing the required capabilities to the coalition capabilities.

2 Calculates the best-expected task outcome of each coalition (coalition weight) and chooses the coalition yielding the best outcome.

4. Issues in Multiple-Robot Systems

The algorithm described in Section 3 yields results that are close to optimal. The current algorithm cannot be directly applied to multiple-robot coalition formation. This section identifies issues that must be addressed for multiple-robot domains.

4.1 Computation vs. Communication

Shehory and Krauss's algorithm (Shehory and Krauss, 1998) requires extensive communication and synchronization during the computation of coalition values. While this may be inexpensive for disembodied agents, it is often desirable to minimize communication in multiple-robot domains even at the expense of extra computation. This work investigates each agent assuming responsibility for all coalitions in which it is a member and thereby eliminating the need for communication. It is necessary to analyze how this would affect each robots computational load. An added assumption is that a robot has a priori knowledge of all robots and their capabilities. Robot capabilities do not typically change; therefore this is not a problem unless a partial or total robot failure is encountered (Ulam and Arkin, 2004). Suppose there are N identical robots and with a perfect computational load distribution, then the number of coalitions each robot must evaluate with communication is:

$$\eta_{with} = \sum_{r=0}^{k} \binom{n}{r} / n \tag{1}$$

The algorithm distributes coalitions between agents as a ratio of their computational capabilities, adding unwanted complexity. It is unlikely that the load will be perfectly distributed, rather some agents will complete their computations before others and remain idle until all computations are completed. The worst case communicational load per agent is $O(n^{k-1})$ during the calculation-distribution stage. If each agent is responsible for only computation of coalitions in which it is a member, then the number of coalitions evaluated with no communication becomes:

$$\eta_{without} = \sum_{r=0}^{k-1} \binom{n-1}{r} \tag{2}$$

Equation 1 requires fewer computations to evaluate but this is not an order of magnitude difference. In both cases, the agent's computational load is O(n^k) per task. The communicational load per robot is O(1) in the calculation-distribution stage. The additional computation may be compensated for by reduced communication time. The Section 5 experiments demonstrate this point. A desirable side effect is additional fault tolerance. If Robot A fails during coalition list evaluation, values for coalitions containing Robot A are lost and those coalitions are no longer considered. Thus a robot failure does not require information retrieval from that robot. However, the other robots must be aware of the failure so that they can delete all coalitions containing the failed robot.

4.2 Task Format

Current multi-agent coalition formation algorithms assume that the agents have a capability vector, $< b_1^i, ..., b_r^i >$. Multiple-robot capabilities include sensors (camera, laser, sonar, or bumper) and actuators (wheels or gripper). Shehory and Krauss's algorithm assumes that the individual agents' resources are collectively available upon coalition formation. The formed coalition freely redistributes resources amongst the members. However, this is not possible in a multiple-robot domain. Robots cannot autonomously exchange capabilities.

Correct resource distribution is also an issue. The box-pushing task can be used to illustrate this point (Gerkey and Mataric, 2002). Three robots cooperate to perform the task, two pushers (one bumper, one camera) and one watcher (one laser, one camera). The total resource requirements are: two bumpers, three cameras, and one laser. However this information is incomplete, as it does not represent the constraints related to sensor locations. Correct task execution requires the laser and camera reside on a single robot. Similarly it is necessary that the bumper and laser reside on different robots. This implies that simply possessing the adequate resources does not necessarily create a multiple-robot coalition that can perform a task, other locational constraints have to be represented and met.

A matrix-based constraint representation is proposed for the multiple-robot domain in order to resolve the problem. The task is represented via a capability matrix called a Task Allocation Matrix (TAM). Each matrix entry corresponds to a capability pair (for example [sonar, laser]). A 1 in an entry indicates that the capability pair must reside on the same robot while a 0 indicates that the pair must reside on separate robots. Finally an X indicates a do not care condition and the pair may or may not reside on the same robot. Every coalition

Table 1. Box-pushing task TAM.

	$Bumper_1$	$Bumper_2$	$Camera_1$	$Camera_2$	$Camera_3$	$Laser_1$
$Bumper_1$	X	0	1	0	0	0
$Bumper_2$	0	X	0	1	0	0
$Camera_1$	1	0	X	0	0	0
$Camera_2$	0	1	0	X	0	0
$Camera_3$	0	0	0	0	X	1
$Laser_1$	0	0	0	0	1	X

must be consistent with the TAM if it is to be evaluated as a candidate coalition. The box-pushing TAM is provided in Table 1. The entry ($Laser_1$, $Camera_3$) is marked 1, indicating that a laser and a camera must reside on the same robot. Similarly the ($Bumper_1$, $Laser_1$) entry is marked 0 indicating the two sensors must reside on different robots.

The TAM can be represented as a Constraint Satisfaction Problem (CSP). The CSP variables are the required sensors and actuators for the task. The domain values for each variable are the available robots possessing the required sensor and actuator capabilities. Two types of constraints exist; the sensors and actuators must reside on the same machine or different machines. A constraint graph can be drawn with locational constraints represented as arcs labeled *s* (same robot) or *d* (different robot). Another constraint is the resource constraint representing that a robot only have as many instances of a sensor and actuator as indicated by the associated capability vector. A robot with one camera can only be assigned one camera node in the constraint graph. Thus all sensors and actuators of the same type have a resource constraint arc labelled *r* between them.

Figure 1 provides the box-pushing task constraint graph. This task's resource constraints between $Bumper_1$ and $Bumper_2$ are implied by their locational constraints. Since $Bumper_1$ and $Bumper_2$ must be assigned to different robots, there cannot be a solution where a robot with one bumper is assigned to both $Bumper_1$ and $Bumper_2$. Similarly the resource constraints between $Camera_1$, $Camera_2$ and $Camera_3$ are implied by the locational constraints between them and there is no need to test them separately. Hence the absence of edges labeled *r*.

The domain values for each variable in the CSP formulation in Figure 1 are the robots that possess the capability represented by the variable. A coalition can be verified to satisfy the constraints by applying arc-consistency. If a sensor is left with an empty domain value set then the current assignment has failed and the current coalition is deemed infeasible. A successful assignment indicates the sub-task to which each robot was assigned.

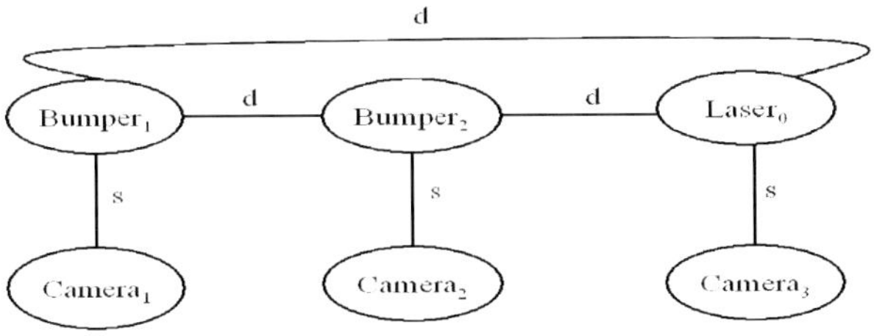

Figure 1. Box-pushing task constraint graph

Using the CSP formulation each candidate coalition is checked to verify if its coalition is feasible. After constraint checking fewer coalitions remain for further evaluation. While additional overhead is incurred during constraint checking, this overhead is somewhat compensated for by the reduced number of coalitions. This is verified by the experimental results in Section 5.

4.3 Coalition Imbalance

Coalition imbalance or lopsidedness is defined as the degree of unevenness of resource contributions made by individual members to the coalition, a characteristic not considered in other coalition formation algorithms. A coalition where one or more agents have a predominant share of the capabilities may have the same utility (coalition weight) as a coalition with evenly distributed capabilities, since robots are unable to redistribute their resources. Therefore coalitions with one or more dominating members (resource contributors) tend to be heavily dependent on those members for task execution. These dominating members then become indispensable. Such coalitions should be avoided in order to improve fault tolerance. Over reliance on dominating members can cause task execution to fail or considerably degrade. If a robot is not a dominating member then it is more likely that another robot with similar capabilities can replace it.

Rejecting lopsided coalitions in favor of balanced ones is not straightforward. When comparing coalitions of different sizes, there can arise a subtle trade-off between lopsidedness and the coalition size. The argument may be made both for fault tolerance and for smaller coalition size. It may be desirable to have coalitions with as few robots as possible. Conversely, there may be a large number of robots thus placing the priority on fault tolerance and balanced coalitions. The Balance Coefficient metric is introduced to quantify the coalition imbalance level. In general, if a coalition has a resource distribution $(r_1, r_2, ..., r_n)$, then the balance coefficient for that coalition with respect to a particular task can be calculated as follows

$$BC = \frac{r_1 \times r_2 \times \ldots r_n}{\left[\frac{taskvalue}{n}\right]^n} \tag{3}$$

A perfectly balanced coalition has a coefficient of 1. The question is how to incorporate the balance coefficient into the algorithm in order to select better coalitions. As previously discussed two cases arise:

1 *Sufficient number of robots and high fault tolerance:* Initially the algorithm proceeds as in Section 3, determining the best-valued coalition without considering lopsidedness. As a modification, a list of all coalitions is maintained whose values are within a certain range (5%) of the best coalition value. The modified algorithm then calculates the balance coefficient for all these coalitions and chooses the most balanced coalition. This ensures that the algorithm always favors the balanced coalition.

2 *Economize on the number of robots:* Maintain a list of all coalitions with values within a bound of the best coalition value. Remove all coalitions larger than the best coalition from the list. Select the coalition with the highest balance coefficient.

5. Experiments

Three experiments testing the validity of the algorithm modifications were conducted, each highlighting a suggested modification. The first experiment measured the variation of time required to evaluate coalitions with and without communication. The number of agents and maximum coalition size were both fixed at five. Communication occurred via TCP/IP sockets over a wireless LAN (see Figure 2). The time for coalition evaluation without communication is significantly less than the time required for evaluation with communication. The time without communication increases at a faster rate as the number of tasks increases. This result occurs because the agent must evaluate a larger number of coalitions when it forgoes communication. Presumably, the two conditions will eventually meet and thereafter the time required with communication will be less than that required without communication. For any practical Agent/Task ratio the time saved by minimizing communication outweighs the extra computation incurred.

The second set of experiments measured the effect of the CSP formulation on the algorithm's execution time. This experiment demonstrates the algorithm's scalability. Figure 3 measures the variation of execution time against the number of agents both with and without constraint checking in the constraint satisfaction graph. Figure 4 shows the variation of execution time compared to the number of tasks. The task complexity in these experiments was similar to the box-pushing task. It can be seen from Figures 3 and 4 that the CSP formulation does not add a great deal to the algorithm's execution or running time. This implies that this formulation can be used to test the validity of a multiple-robot coalition without incurring much overhead.

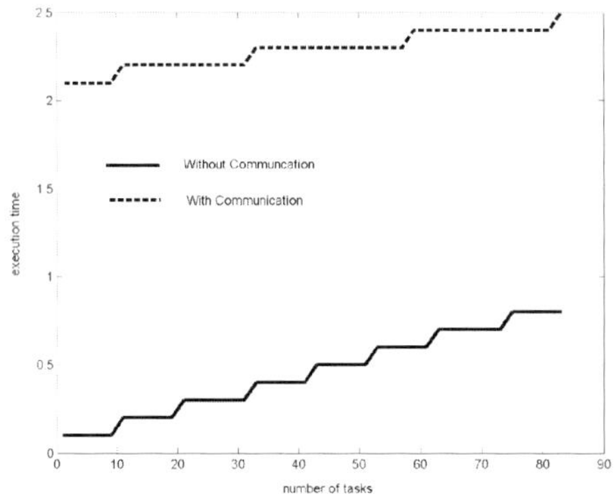

Figure 2. Execution time with and without communication

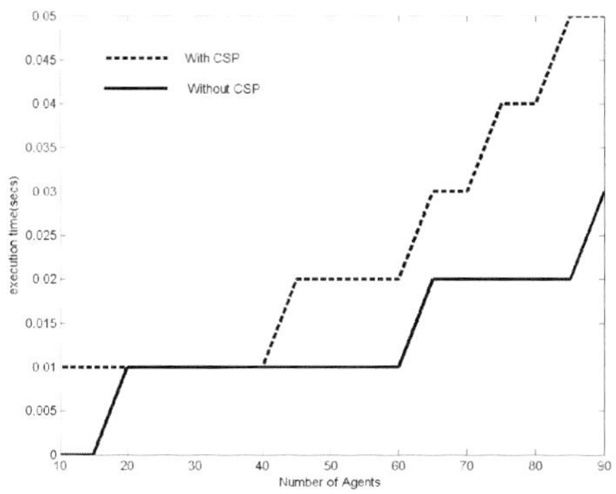

Figure 3. Execution time vs. Number of Agents

The third set of experiments demonstrates the effect of utilizing the Balance Coefficient to favor the creation of balanced coalitions. The Player/Stage simulation environment(Gerkey et al., 2003) was employed for this experiment. The simple tasks required breaking up a formation of resized hockey pucks by bumping into the formation. The degree of task difficulty was adjusted by varying the hockey pucks' coefficient of friction with the floor. Adjusting the forces they could exert varied the robots' capabilities. There are no locational constraints on the task capability requirements. Ten simulated robots were used for the experiment, as shown in Figure 5. The robots were numbered 1 to 10 from the top of the figure along left side. Each robot had a specific capability type: small robots had 10 units of force (robots 1, 2, 8, 9, 10), medium sized robots had 15 units of force (robots 5, 6, 7) and large robots had 20 units of force (robots 3, 4). Simulation snapshots are provided for a task requiring

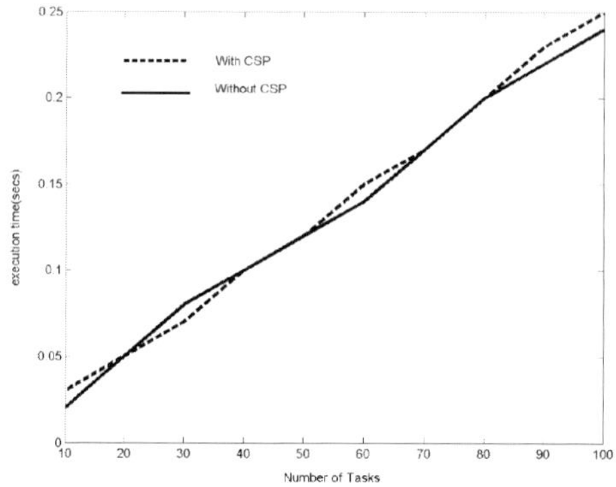

Figure 4. Execution time vs. Number of Tasks

Figure 5. Two large robots and one small robot form a coalition

50 units of force. Figure 5 shows the formed coalition without balancing. The coalition is comprised of two large robots and one small robot.

Figure 6 shows the same task incorporating the balance coefficient into the coalition formation. This choice places a low priority on fault tolerance and a high priority on economizing the number of robots (Case 1 from Section 4.3). The formed coalition is comprised of two medium sized robots and one large robot. The resulting coalition is more balanced and has a higher balance coefficient (0.972 as opposed to 0.864 for the coalition in Figure 6). Figure 7 depicts the experiment conducted with no restrictions on the coalition size (Case 2 from Section 4.3). The resulting coalition consists of five small robots. Thus a perfectly balanced coalition (balance coefficient = 1) is obtained when the coalition size is unconstrained. The advantage is that a larger number of small (less capable) robots should have higher fault tolerance. If one robot fails, it should be easier to find a replacement as opposed to replacing a larger (more capable) robot.

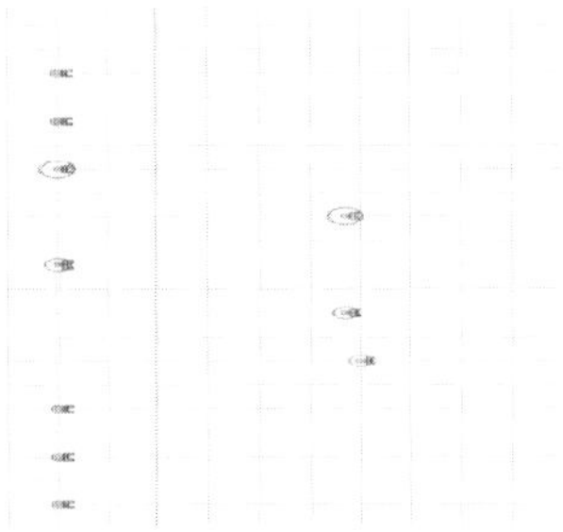

Figure 6. One large and two medium sized robots form a coalition

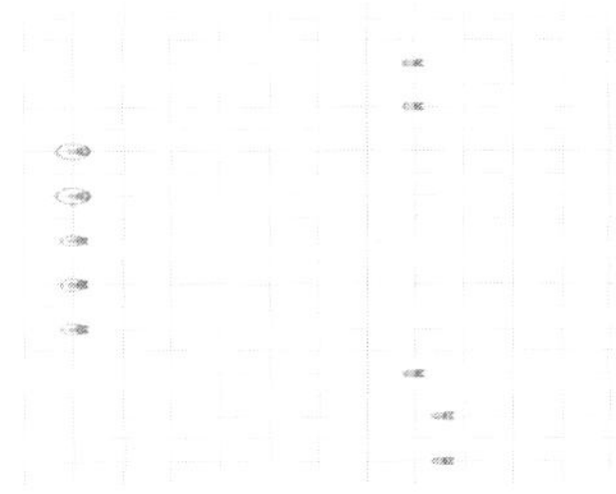

Figure 7. Five small robots form a coalition

6. Conclusion and Future Work

Finding the optimal multiple-robot coalition for a task is an intractable problem. This work shows that, with certain modifications, coalition formation algorithms provided in the multi-agent domain can be applied to the multiple-robot domain. This paper identifies modifications and incorporates them into an existing multi-agent coalition formation algorithm. The impact of extensive communication between robots was shown to be severe enough to endorse relinquishing communication in favor of additional computation when possible. The task format in multi-robot coalitions was modified to adequately represent additional constraints imposed by the multiple-robot domain. The concept of coalition imbalance was introduced and its impact on the coalition's fault tolerance was demonstrated.

Further algorithm modifications will permit more complex task execution by utilizing a MURDOCH (Gerkey and Mataric, 2002) style task allocation scheme within coalitions. A future goal is to investigate methods of forming coalitions within a dynamic real-time environment. The long-term goal is to develop a highly adaptive, fault tolerant system that would be able to flexibly handle different tasks and task environments.

References

Anderson, J. E., Tanner, B., and Baltes, J. (2004). Dynamic coalition formation in robotic soccer. Technical Report WS-04-06, AAAI workshop.

Brooks, C. H. and Durfee, E. H. (2003). Congregation formation in multiagent systems. *Autonomous Agents and Multi-Agent Systems*, 79:145–170.

Fass, L. (2004). An automatic-theoretic view of agent coalitions. Technical Report WS-04-06, AAAI workshop.

Gerkey, B. and Mataric, M. (2002). Sold! auction methods for multirobot coordination. *IEEE Transactions on Robotics and Automation*, 18:758–68.

Gerkey, B. and Mataric, M. (2004). A framework for studying multi-robot task allocation. *International Journal of Robotics Research*. to appear.

Gerkey, B., Vaughan, R. T., and Howard, A. (2003). The player/stage project: Tools for multi-robot and distributed sensor systems. In *11th Intr. Conf. on Advanced Robotics*, pages 317–323.

Li, X. and Soh, L.-K. (2004). Investigating reinforcement learning in multiagent coalition formation. Technical Report WS-04-06, AAAI workshop.

Parker, L. (1998). Alliance: An architecture for fault tolerant multi-robot cooperation. *IEEE Transactions on Robotics and Automation*, 14:220–240.

Sandholm, T., Larson, K., Andersson, M., Shehory, O., and Tomhe, F. (1999). Coalition structure generation with worst case guarantees. *Artificial Intelligence*, 111:209–238.

Shehory, O. and Krauss, S. (1996). A kernel oriented model for coalition-formation in general environments: Implementation and results. In *AAAI*, pages 134–140.

Shehory, O. and Krauss, S. (1998). Methods for task allocation via agent coalition formation. *Artificial Intelligence Journal*, 101:165–200.

Shehory, O. and Krauss, S. (1999). Feasible formation of coalitions among autonomous agents in non-super-additive environments. *Computational Intelligence*, 15:218–251.

Sorbella, R., Chella, A., and Arkin, R. (2004). Metaphor of politics: A mechanism of coalition formation. Technical Report WS-04-06, AAAI workshop.

Ulam, P. and Arkin, R. (2004). When good comms go bad: Communcations recovery for multi-robot teams. In *2004 IEEE Intr. Conf. on Robotics and Automation*, pages 3727–3734.

Werger, B. and Mataric, M. (2000). Broadcast of local eligibility: Behavior-based control for strongly-cooperative robot teams. In *Autonomous Agents*, pages 347–356.

SENSOR NETWORK-MEDIATED MULTI-ROBOT TASK ALLOCATION

Maxim A. Batalin and Gaurav S. Sukhatme
Robotic Embedded Systems Laboratory
Center for Robotics and Embedded Systems
Computer Science Department
University of Southern California
Los Angeles, CA 90089, USA
maxim@robotics.usc.edu, gaurav@usc.edu

Abstract We address the *Online Multi-Robot Task Allocation (OMRTA)* problem. Our approach relies on a computational and sensing fabric of networked sensors embedded into the environment. This sensor network acts as a distributed sensor and computational platform which computes a solution to OMRTA and directs robots to the vicinity of tasks. We term this Distributed In-Network Task Allocation (DINTA). We describe DINTA, and show its application to multi-robot task allocation in simulation, laboratory, and field settings. We establish that such network-mediated task allocation scales well, and is especially amendable to simple, heterogeneous robots.

Keywords: Mobile robots, sensor networks, task allocation, distributed

1. Introduction

We focus on the intentional cooperation of robots toward a goal ((Parker, 1998)). Within such a setting, a natural question is the assignment of robots to sub-goals such that the ensemble of robots achieves the overall objective. Following ((Gerkey and Matarict', 2004)) we call such sub-goals, *tasks*, and their assignment to robots, the *Multi-Robot Task Allocation (MRTA)* problem. Simply stated, MRTA is a problem of assigning or allocating tasks to (intentionally cooperating) robots over time such that some measure of overall performance is maximized.

We focus on the online version of the problem (OMRTA), where 1. tasks are geographically and temporally spread, 2. a task schedule is not available in advance, and 3. robots need to physically visit task locations to accomplish task completion (*e.g.,* to push an object). Our approach to OMRTA relies on a computational and sensing fabric of networked sensors embedded into the

L.E. Parker et al. (eds.),
Multi-Robot Systems. From Swarms to Intelligent Automata. Volume III, 27–38.

environment. This sensor network acts as a distributed sensor and computational platform which computes a solution to OMRTA and directs robots to the vicinity of tasks. To make a loose analogy, robots are routed from source to destination locations in much the same way packets are routed in conventional networks. We term this, Distributed In-network Task Allocation (DINTA).

There are five advantages to doing the task allocation in this manner:

1. **Simplicity:** Since the task-allocation is done in the network, robots may be very simple, designed specifically for optimal task *execution* (*e.g.,* specialized end effectors) rather than computational sophistication. Further, robots do not need conventional localization or mapping support.

2. **Communication:** Robots are not required to be within communication range of each other. The network is used for propagating messages between the robots.

3. **Scaling:** There is no computation or communication overhead associated with increasing the number of robots.

4. **Identity:** Robots are not required to recognize each other.

5. **Heterogeneity:** Robots may be of different types, and need only a common interface to the sensor network.

In this paper we make the following contributions. We briefly review the details of DINTA[1], and demonstrate its application to a system for spatiotemporal monitoring of environmental variables in nature. We note that while we study the task allocation problem in the context of mobile robots, sensor network-mediated task allocation can also be used in other settings (*e.g.,* in an emergency people trying to leave a building would be guided (tasked) to the closest exits by the network).

2. Related Work

The problem of multi-robot task allocation (MRTA) has received considerable attention. For an overview and comparison of the key MRTA architectures see ((Gerkey and Matarict', 2004)), which subdivides MRTA architectures into behavior-based and auction-based. For example, ALLIANCE ((Parker, 1998)) is a behavior-based architecture that considers all tasks for (re)assignment at every iteration based on robots' utility. Utility is computed by measures of acquiescence and impatience. Broadcast of Local Eligibility ((Werger and Matarit'c, 2000)) is also a behavior-based approach, with fixed-priority tasks. For every task there exists a behavior capable of executing the task and estimating the utility of robot executing the task. Auction-based approaches include the M+ system ((Botelho and Alami, 2000)) and Murdoch ((Gerkey and

Mataricť, 2004)). Both systems rely on the Contract Net Protocol (CNP) that makes tasks available for auction, and candidate robots make 'bids' that are their task-specific utility estimates. The highest bidder (i.e., the best-fit robot) wins a contract for the task and proceeds to execute it. All previous MRTA approaches in the robotics community have focused on performing the task allocation computation on the robots, or at some centralized location external to the robots. All the sensing associated with tasks, and robot localization, is typically performed on the robots themselves. Our approach relies on a sensor network, which performs event detection and task-allocation computation, allowing robots to be simple and heterogeneous.

3. Distributed In-Network Task Allocation: DINTA

As an experimental substrate, we use a particular stylized monitoring scenario in which robots are tasked with 'attending' to the environment such that areas of the environment in which something significant happens, do not stay unattended for long. We model this using the notion of alarms. An alarm is spatially focused, but has temporal extent (*i.e.,* it remains on until it is turned off by a robot). Alarms are detected by sensor nodes embedded in the environment. For example in a natural setting, an alarm might be generated in case an abrupt change in temperature is detected requiring inspection of the area by the robot. The task of the team of robots is to turn off the alarms by responding to each alarm. This is done by a robot navigating to the location of the alarm. Once the robot arrives in the vicinity of the alarm, the alarm is deactivated. Thus the robot response is purely notional in that the task the robot performs is to arrive at the appropriate location only. The goal is to minimize the cumulative alarm *On Time* across all alarms, over the duration of the entire experimental trial. Each alarm's *On Time* is computed as the difference between the time the alarm was deactivated by a robot and the time the alarm was detected by one of the nodes of the network.

The basic idea of DINTA is that given a set of alarms (each corresponding to a task) detected by the network (*e.g.,* nodes detect motion, presence of dangerous chemicals, etc.), every node in the network computes a suggested 'best' motion direction for all robots in its vicinity. The ensemble of suggested directions computed over all nodes is called a navigation field. In case multiple tasks arrive at the same time, multiple navigation fields (one for every task) are maintained in the network and explicitly assigned to robots. Navigation fields are assigned to robots using a greedy policy.

3.1 Computing Navigation Field

We assume that the network is deployed and every node stores a discrete probability distribution of the transition probability $P(s'|s_C,a)$ (probability of

Algorithm 1. Adaptive Distributed Navigation Field Computation Algorithm (running on every node).

s - current node
S - set of all nodes
$A(s)$ - set of all actions possible from node s
$C(s,a)$ - cost of taking an action a from node s
$P(s'|s,a)$ - probability of arriving at node s' given
that the robot started at node s and commanded an action a,
stored on node s
$\pi(s)$ - optimal direction that robot should take at node s

Compute Direction(goal_node)
if $s ==$ goal_node
  $V_0 =$ some big number
else
  $V_0 = 0$
while $V_t - V_{t-1} > \varepsilon$ **do**
  Query neighbor nodes for their new values V_t
  if received new values V_t from all neighbor nodes s'
    $V_{t+1}(s) = C(s,a) + \max_{a \in A(s)} \sum_{s' \in S-s} P(s'|s,a) \times V_t(s')$
    Update neighbor nodes with new value $V_{t+1}(s)$
Query neighbor nodes for their final values $V(s')$
$\pi(s) = \arg\max_{a \in A(s)} \sum_{s' \in S-s} P(s'|s,a) \times V(s')$

the robot arriving at node s' given that it started at node s_C and was told to execute action a). The reader is referred to ((Batalin and Sukhatme, 2004a)) for a detailed discussion on how such distributions can be obtained.

Algorithm 1 shows the pseudo code of the adaptive distributed navigation field computation algorithm, which runs on every network node. We use value iteration ((Koenig and Simmons, 1992)) to compute the best action at a given node. The general idea behind value iteration is to compute the values (or utilities) for every node and then pick the actions that yield a path towards the goal with maximum expected value. Expected values are initialized to 0. Since $C(s,a)$ is the cost associated with moving to the next node, it is chosen to be a negative number which is smaller than $\frac{-(minimal\_reward)}{k}$, where k is the number of nodes. The rationale is that the robot should pay for taking an action (otherwise any path the robot might take would have the same value), however, the cost should not be too large (otherwise the robot might prefer to stay at the same node).

Next, as shown in Algorithm 1, a node queries its neighbors for the latest utility values V. Once the values are obtained from all neighbors, a node updates its own utility. This process continues until the values do not change beyond an ε (set to 10^{-3} in our experiments). After the latest values from all neighbors are collected, a node can compute an action policy π (optimal direction) that a robot should take if it is in the node's vicinity.

In combination, the optimal directions computed by individual network nodes, constitute a global navigation field. Practical considerations for robot navigation using this approach are discussed in ((Batalin et al., 2004b)).

3.2 Task Allocation

DINTA assigns tasks in *decision epochs* - short intervals of time during which only the tasks that have arrived since the end of the previous epoch are considered for assignment. The following describes the behavior of DINTA in a particular epoch e. Let the network detect two alarms A_1 and A_2 (Figure 1a) by nodes a_1 and a_2 respectively in an epoch e. Both nodes a_1 and a_2 notify the entire network about the new alarms and start two navigation field computations (using Algorithm 1) - one for each goal node. Next consider nodes r_1 and r_2 that have unassigned robots R_1 and R_2 (Figure 1b) in their vicinity. r_1 and r_2 propagate the distances between the unassigned robots and the alarms A_1 and A_2. Four such distances are computed and distributed throughout the network. In the final stage, every node in the network has the same information about the location of alarms and available robots, and distances between the robots and each alarm. Each node in the network can now decide uniquely which navigation field to assign to which robot. Figure 1c shows two navigation fields (one for each robot) generated and assigned to the robots. A robot then simply follows the directions suggested by network nodes.

4. MRTA Experiments in Simulation

In the first set of experiments described here we used the Player/Stage ((Gerkey et al., 2003)) simulation engine populated with simulated Pioneer 2DX mobile robots. A network of 25 network nodes (simulated motes ((Pister et al., 1999))) was pre-deployed in a test environment of size $576m^2$. The communication range of the nodes and robots was set to approximately 4 meters. Robots were required to navigate to the point of each alarm and minimize the cumulative alarm *On Time*. Each alarm's *On Time* is computed as the difference between the time the alarm was served by a robot and the time the alarm was detected by one of the nodes of the sensor network. Every experiment was conducted in the same environment with robot group sizes varying from 1 to 4, 10 trials per group. The schedule of 10 alarms was drawn from a Poisson distribution ($\lambda = \frac{1}{60}$, roughly one alarm per minute), with uniformly distributed nodes that detected alarms.

We measured cumulative alarm *On Time* for network-mediated task allocation (*i.e.,* DINTA). As a base case we compared the results to the situation where the robots are programmed to explore the environment using directives

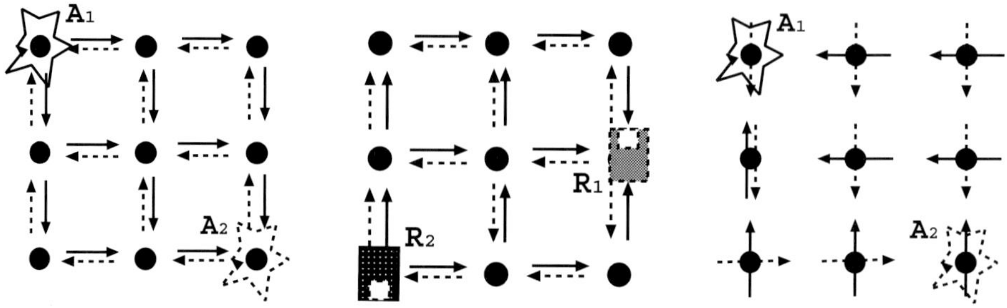

(a) Phase 1. (b) Phase 2. (c) Phase 3.

Figure 1. The three stages of DINTA in a decision epoch. a) The sensor network detects events (marked A_1 and A_2) and propagates event data throughout the network. b) Next, nodes that have unassigned robots in their vicinity propagate distances (in hop counts) from robots to each of the alarms. c) In the final stage, every node in the network has the same information about the location of events and available robots, and distances between robots and each event. Hence, a unique assignment of direction suggestion at every node can occur.

from the sensor network designed only to optimize their environmental coverage ((Batalin and Sukhatme, 2004a)). The comparison highlights the benefits of purposeful task allocation. Figure 2 shows the *OnTime* comparison for DINTA and the exploration-only case. Clearly, DINTA outperforms the exploration-only algorithm even though as the environment becomes saturated with robots, the difference becomes smaller. The difference is statistically significant (the T-test p-value is less than 10^{-4} for every pair in the data set). Further, the performance of DINTA is stable (small and constant variance) whereas variances produced by the exploration-only mode change drastically and reduce as the environment becomes saturated with robots.

5. Laboratory Experiments with NIMS

The second set of experiments we discuss use a new testbed, currently under development - Networked Info-Mechanical System ((NIMS, 2004)). Figure 3 shows NIMS deployed in a forest reserve for continuous operation. The system includes supporting cable infrastructure, a horizontally moving mobile robot (the NIMS node) equipped with a camera, and a vertically mobile meteorological sensor system carrying water vapor, temperature, and photosynthetically active radiation (PAR) sensing capability. The purpose of NIMS is to enable the study of spatiotemporal phenomena (*e.g.,* humidity, carbon flux, *etc.*) in natural environments. Figure 3a schematically shows NIMS with deployed static sensor nodes (assembled in strands) in the volume surrounding the sensing

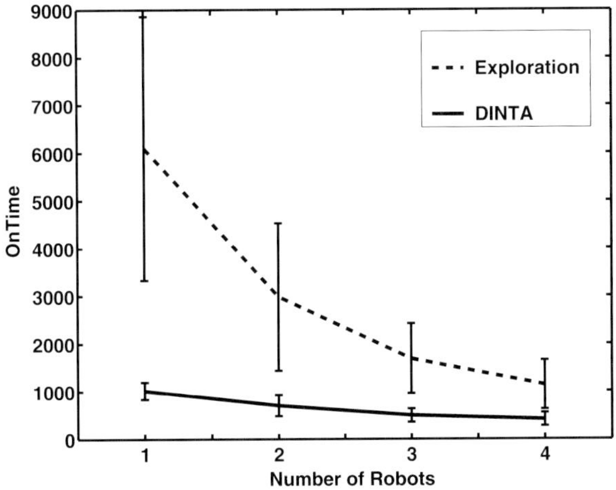

Figure 2. Comparison between implementation of DINTA and exploration-only.

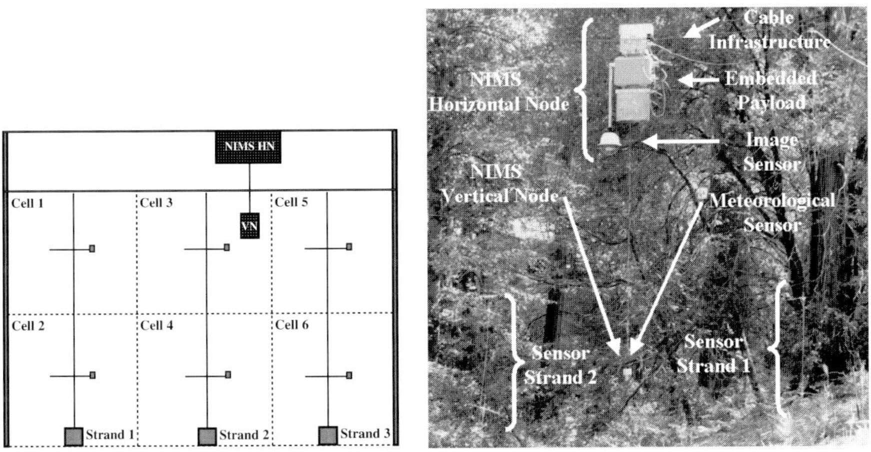

(a) NIMS horizontal (HN) and vertical (VN) nodes and static sensors (schematically)

(b) NIMS deployed in a forest reserve

Figure 3. NIMS system deployed in the forest reserve for continuous operation.

transect. Wireless networking is incorporated to link the static sensor nodes with the NIMS node. The NIMS system is deployed in a transect of length 70m and average height of 15m with a total area of over 1,000 m^2.

The experimental NIMS system operates with a linear speed range for node motion of 0.1 to 1 m/second. Thus, the time required to map an entire 1,000 m^2 transect with 0.1 m^2 resolution will exceed 10^4 to 10^5 seconds. Phenom-

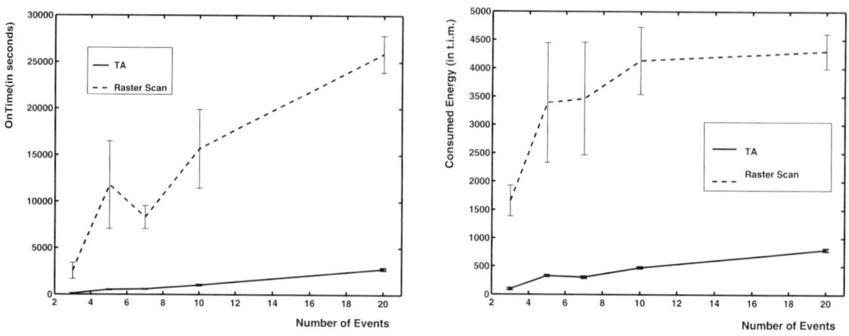

(a) Event OnTime. (b) Energy consumption.

Figure 4. NIMS lab experiments: task allocation vs. a raster scan.

ena that vary at a characteristic rate exceeding this scanning rate may not be accurately represented. Hence task allocation is required to focus sampling in specific areas depending on their scientific value. The preliminary experiments using our in-network task allocation methodology show an order of magnitude improvement in the time it takes to complete sampling.

We conducted experiments on a smaller version of NIMS installed in the lab[2]. A network of 6 Mica2 motes was pre-deployed in the volume surrounding the NIMS transect (similar to Figure 3a) in a test environment. Experiments were conducted comparing a version of DINTA with a Raster Scan (RS) as a base case. RS is an algorithm of choice when there is no information about the phenomenon location (where the alarms are). RS scans every point of the transect with a specified resolution. When the Raster Scan reaches the location of an alarm, the alarm is considered to be turned off.

In our experiment, schedules of 3, 5, 7, 10 and 20 alarms (henceforth, events) were drawn from a uniform distribution to arrive within 10 minutes, with uniformly distributed nodes that detected the event. Note that for actual applications we do not expect to receive/process more than 1 - 10 events in 10 minutes on average. Hence the case of 20 events shows the behavior of the system at the limit.

Figure 4 shows experimental results comparing *OnTime* performance of DINTA and RS. The number of events varies between 3 and 20. Both algorithms were evaluated from 3 different starting positions of the mobile node on the transect (drawn from a uniform distribution). The results were averaged. As can be seen from the graph, DINTA performs 9-22 times better on the entire interval of 3-20 events. Note also that DINTA is stable, as indicated by error bars, and hence is favored for use in this application since it provides reduced bounds on system run time over a simple Raster Scan method.

We also compared mobility requirements for DINTA and RS methods. Specifically, the use of mobility requires energy. A measure of energy for mobility is determined for the purposes of comparison by computing the total time of the robot motion. Figure 4 shows a comparison of energy consumption in units of time-in-motion. As expected, DINTA outperforms Raster Scan significantly. However as the number of events increases to infinity, DINTA will approach Raster Scan energy consumption. Also note, that on the interval [5,20] the slope of the Raster Scan curve is very small and the energy consumption is insensitive to event arrival rate.

6. Field Trials using NIMS

The third, and final, set of experiments discussed here were performed in field trials with the NIMS system. We used our task allocation system and compared two policies - *Time* (tasks with smaller time stamp get priority) and *Distance* (tasks closer to the robot get priority). A set of experiments was conducted on a NIMS setup deployed in the James San Jacinto Mountain Reserve. Because of space limitations, only representative graphs are presented. Figure 5 shows the representative PAR data from sensor 1 collected during the operation of the *Time* policy (Figure 5a) and the *Distance* policy (Figure 5b). Figure 5 also shows points in time when events were generated and serviced by both policies for sensor 1. Note that events are generated in response to fluctuations in PAR. As shown on Figure 5, events are generated proportionally to the density of the 'spikes' in PAR data and cover all significant 'spikes' of PAR data.

Figure 5c shows the comparison between the cumulative event *OnTime* of the *Time* policy and the *Distance* policy. For visualization purposes, in Figure 5c event's *OnTime* is presented as a zero-mean Gaussian distribution. It follows that the *Distance* policy has smaller average *OnTime* with smaller deviation.

7. Summary

We presented a novel, sensor network-mediated, approach to multi robot task allocation. Our algorithm DINTA: Distributed In-Network Task Allocation solves the online multi robot task allocation problem. This approach allows us to combine the benefits of a sensor network with the mobility and functionality of robots. The system computes task assignments distributively in-network while, at the same time, providing a virtual sensor and communication device that 'extends' throughout the whole environment. There are several advantages in using DINTA as opposed to traditional MRTA approaches. The

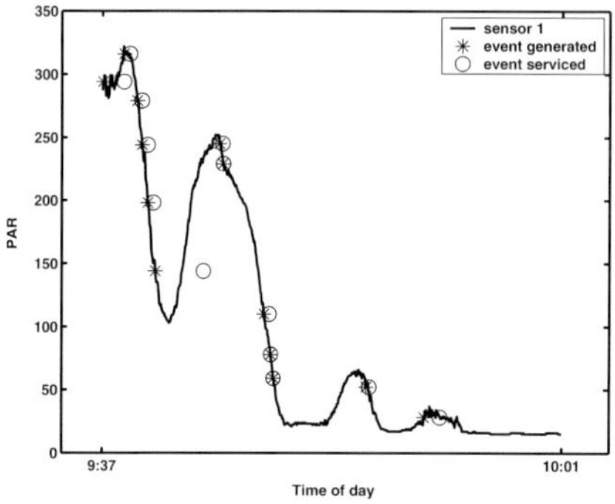

(a) Time policy.

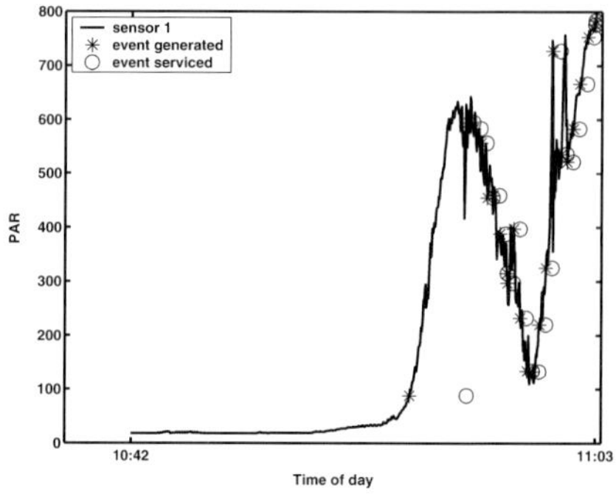

(b) Distance policy.

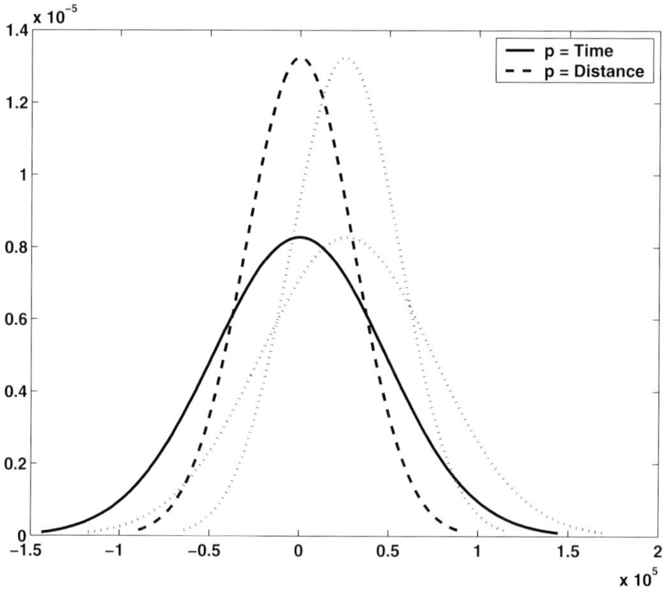

(c) Event OnTime.

Figure 5. NIMS field experiments for two policies. a,b) PAR data acquired by the first sensor during one of the field experiments. Events generated and serviced are shown for Time and Distance policies. Note that events are rendered time of occurrence vs. the PAR value of the event. c) Event OnTime in a form of a zero-mean Gaussian distributions for Time and Distance policies. The OnTime of events generated by all 6 sensors is considered. Dotted (blue or lighter) graphs show the distributions at original means.

sensor network allows a robot to detect a goal (alarm, event) even though the alarm is not in the robot's sensor range. In addition, robots can use the sensor network to relay messages if they are not within communication range of each other. Further, robots can be very simple and potentially heterogeneous. We also presented physical experimental results of using DINTA for field measurements in natural setting using a monitoring infrastructure composed of mobile robots on cables and network nodes in the vicinity of the cable transect.

Acknowledgments

This work is supported in part by the National Science Foundation (NSF) under grants IIS-0133947, EIA-0121141 and grants CCR-0120778, ANI-00331481 (via subcontract). Any opinions, findings, and conclusions or recommendations expressed in this material are those of the authors and do not necessarily reflect the views of the National Science Foundation.

Notes

1. For implementation details of DINTA see ((Batalin and Sukhatme, 2004b).)
2. For experimental and other details see ((Batalin et al., 2004a)).

References

Batalin, M., Rahimi, M., Yu, Y., Liu, S., Kansal, A., Sukhatme, G., Kaiser, W., Hansen, M., Pottie, G., Srivastava, M., and Estrin, D. (2004a). Call and Response: Experiments in Sampling the Environment. In *Proc. ACM SenSys*.

Batalin, M. and Sukhatme, G. (2004a). Coverage, Exploration and Deployment by a Mobile Robot and Communication Network. *Telecommunication Systems Journal, Special Issue on Wireless Sensor Networks*, 26(2):181–196.

Batalin, M. and Sukhatme, G. (2004b). Using a Sensor Network for Distributed Multi-Robot Task Allocation. In *Proc. IEEE International Conference on Robotics and Automation*, pages 158–164, New Orleans, Louisiana.

Batalin, M., Sukhatme, G., and Hattig, M. (2004b). Mobile Robot Navigation using a Sensor Network. In *Proc. IEEE International Conference on Robotics and Automation*, pages 636–642, New Orleans, Louisiana.

Botelho, S. and Alami, R. (2000). M+: A Scheme for Multi-Robot Cooperation through Negotiated Task Allocation and Achievement. In *Proc. IEEE International Conference on Robotics and Automation (ICRA)*, pages 293–298.

Gerkey, B., Vaughan, R., and Howard, A. (2003). The Player/Stage Project: Tools for Multi-Robot and Distributed Sensor Systems. In *Proc. International Conference on Advanced Robotics (ICAR 2003)*, pages 317–323, Coimbra, Portugal.

Gerkey, B. P. and Matarict', M. J. (2004). A Formal Analysis and Taxonomy of Task Allocation in Multi-Robot Systems. *Intl. Journal of Robotics Research*, 23(9):939–954.

Koenig, S. and Simmons, R. G. (1992). Complexity Analysis of Real-Time Reinforcement Learning Applied to Finding Shortest Paths in Deterministic Domains. Technical Report CMU-CS-93-106, Carnegie Mellon University, School of Computer Science, Carnegie Mellon University, Pittsburg, PA 15213.

NIMS (2004). http://www.cens.ucla.edu/portal/nims.html.

Parker, L. E. (1998). ALLIANCE: An Architecture for Fault-Tolerant Multi-Robot Cooperation. In *IEEE Transactions on Robotics and Automation*, volume 14, pages 220–240.

Pister, K. S. J., Kahn, J. M., and Boser, B. E. (1999). Smart Dust: Wireless Networks of Millimeter-Scale Sensor Nodes. *Electronics Research Laboratory Research Summary*.

Werger, B. B. and Matarit'c, M. J. (2000). *Distributed Autonomous Robotic Systems 4*, chapter Broadcast of Local Eligibility for Multi-Target Observation, pages 347–356. Springer-Verlag.

II

COORDINATION IN DYNAMIC ENVIRONMENTS

MULTI-OBJECTIVE COOPERATIVE CONTROL OF DYNAMICAL SYSTEMS

Zhihua Qu
Department of Electrical and Computer Engineering, University of Central Florida, Orlando, FL 32816, USA *
qu@mail.ucf.edu

Jing Wang
Department of Electrical and Computer Engineering, University of Central Florida, Orlando, FL 32816, USA
jwang@pegasus.cc.ucf.edu

Richard A. Hull
Lockheed Martin Missiles and Fire Control, 5600 Sand Lake Road, Orlando, FL 32819, USA
Richard.A.Hull@lmco.com

Abstract In this paper, the cooperative control problem of making system's different outputs converge to different steady states is studied for a general class of MIMO dynamic systems with finite but arbitrary relative degree. A set of less-restrictive conditions on the design of cooperative control feedback matrices are established. In particular, the proposed design does not require either that collaborative robots have a fixed communication/control structure (such as leader/followr or nearest neighbor) or that their sensor/communication graph be irreducible.

Keywords: Cooperative control, dynamical systems, multi-objective

1. Introduction

Recently, the cooperative control problem of making the states of a group of dynamical systems converge to same steady state has stirred a great deal of interests since its solvability implies the solvability of general formation stabilization problem which has direct applications in many fields requiring

*The research is supported in part by a grant from Lockheed Martin Corporation.

L.E. Parker et al. (eds.), Multi-Robot Systems. From Swarms to Intelligent Automata. Volume III, 41–52.

multiple inexpensive mobile systems, such as deploying a group of vehicles for search or exploration purpose in hazardous environments.

Motivated by the flocking behavior of birds in flight, Reynolds introduced a computer animation model for cohesion, separation and alignment in (Reynolds, 1987). Subsequently, a simple discrete-time model (Vicsek model) was given in (Vicsek et al., 1995) for the heading alignment of autonomous particles moving in the plane. Simulation results verified the correctness of Vicsek model. More recently, a theoretical explanation of Vicsek model was presented in (Jadbabaie et al., 2003) by using graph theory. The conditions on the connectivity of undirected sensor graphs are given for the overall system convergence. This result was extended to networks with directed sensor graphs in (Lin et al., 2004)(Moreau, 2003). Other recent relevant works include the consensus problems in (Saber and Murray, 2003)(Ren and Beard, 2004). Most of the aforementioned works only considered simple linear systems with single or double integrator models. Moreover, for multiple output system, it is desirable in practice that the system's different outputs converge to different steady states.

In our recent works (Qu et al., 2004a)(Qu et al., 2004b), we extend the results in (Jadbabaie et al., 2003)(Lin et al., 2004) to a general class of MIMO dynamical systems of finite but arbitrary relative degree. In particular, a leader-follower cooperative control strategy was analyzed in (Qu et al., 2004a) and general leaderless cooperative control was studied in (Qu et al., 2004b). In this paper, we continue the works in (Qu et al., 2004a)(Qu et al., 2004b), and study the problem of making systems' different channels (outputs) as well as the corresponding states converge to different steady states. The proposed design does not require either that collaborative robots have a fixed communication/control structure (such as leader/followr or nearest neighbor) or that their sensor/communication graph be irreducible. The convergence of the overall system is rigorously proved by studying the convergence of products of a sequence of row stochastic matrices. An illustrative example is provided to verify the proposed method.

2. Preliminaries

Let $\mathbf{1}_p$ be the p-dimensional column vector with all its elements being 1, and $\mathbf{J}_{r_1\times r_2} \in \Re^{r_1\times r_2}$ be a matrix whose elements are all 1. I_m is the $m-$dimensional identity matrix. $\otimes$ denotes the Kronecker product. A nonnegative matrix has all entries nonnegative. A square real matrix is row stochastic if it is nonnegative and its row sums all equal 1. For a row stochastic matrix E, define $\delta(E) = \max_j \max_{i_1,i_2} |E_{i_1 j} - E_{i_2 j}|$, which measures how different the rows of E are. Also, define $\lambda(E) = 1 - \min_{i_1,i_2} \sum_j \min(E_{i_1 j}, E_{i_2 j})$. Given a sequence of nonnegative matrix $E(k)$, the notation of $E(k) \succ 0, k = 0, 1, \cdots,$ means that, there is a sub-sequence $\{l_v, v = 1, \cdots, \infty\}$ of $\{0, 1, 2, \cdots, \infty\}$ such that $\lim_{v \to \infty} l_v = +\infty$ and $E(l_v) \neq 0$. A non-negative matrix E is said to be *reducible* if the set of its indices, $\mathcal{I} \triangleq \{1, 2, \cdots, n\}$, can be divided into two disjoint non-empty sets $\mathcal{S} \triangleq \{i_1, i_2, \cdots, i_\mu\}$ and $\mathcal{S}^c \triangleq \mathcal{I}/\mathcal{S} = \{j_1, j_2, \cdots, j_\nu\}$ (with $\mu + \nu = n$) such that $E_{i_\alpha j_\beta} = 0$, where $\alpha = 1, \cdots, \mu$ and $\beta = 1, \cdots, \nu$. Matrix E is said to be *irreducible* if it is not reducible. A square matrix $E \in \Re^{n \times n}$ can be used to define a directed graph with n nodes $v_1, \cdots, v_n$, and there is a directed arc from v_i to v_j if and only if $E_{ij} \neq 0$. A directed graph represented by E is *strongly connected* if between every pair of distinct nodes v_i, v_j there is a directed path of finite length that begins at v_i and ends at v_j. The fact that a directed graph represented by E is strongly connected is equivalent to that matrix E is irreducible (Minc, 1988).

The following lemma provides a necessary and sufficient condition on the irreducibility of a non-negative matrix in a special structure, which has direct relation to the canonical system model discussed in this paper.

Lemma 2.1. *(Qu et al., 2004a) Given any non-negative matrix $E \in \Re^{(qm) \times (qm)}$ with sub-blocks $E_{ij} \in \Re^{m \times m}$, let $\overline{E} = [\overline{E}_{ij}] \in \Re^{(Lm) \times (Lm)}$ with $L = l_1 + \cdots + l_q$ be defined by*

$$\overline{E}_{ii} = \begin{bmatrix} 0 & I_{(l_i - 1)} \otimes I_m \\ E_{ii} & 0 \end{bmatrix}, \quad \overline{E}_{ij} = \begin{bmatrix} 0 & 0 \\ E_{ij} & 0 \end{bmatrix}$$

where $l_i \geq 1$ are positive integers for $i = 1, \cdots, q$. Then, $\overline{E}$ is irreducible if and only if E is irreducible.

The classical convergence result of the infinite products of sequences of row stochastic matrices (Wolfowitz, 1963) has been applied in the study of coordination behavior of groups of mobile autonomous agents (Jadbabaie et al., 2003)(Lin et al., 2004). In our recent works (Qu et al., 2004a)(Qu et al., 2004b), we relaxed the condition in (Wolfowitz, 1963) and found an easy-to-check condition on the convergence of a sequence of row stochastic matrices in the lower-triangular structure, and also extended it to the case of the products of lower-triangular matrices and general matrices. These new results are useful for establishing less-restrictive conditions on the design of cooperative control and the connectivity requirements among individual systems. In what follows, we recall these two results without proof for briefness.

Lemma 2.2. *(Qu et al., 2004a) Consider a sequence of nonnegative, row stochastic matrices $P(k) \in \Re^{R \times R}$ in the lower-triangular structure, where $R = \sum_{i=1}^{m} r_i$, sub-blocks $P_{ii}(k)$ on the diagonal are square and of dimension $\Re^{r_i \times r_i}$, sub-blocks $P_{ij}(k)$ off diagonal are of appropriate dimensions. Suppose that $P_{ii}(k) \geq \varepsilon_i \mathbf{J}_{r_i \times r_i}$ for some constant $\varepsilon_i > 0$ and for all $(i = 1, \cdots, m)$, and in the ith row of $P(k)$ $(i > 1)$, there is at least one j $(j < i)$ such that $P_{ij} \succ 0$. Then,*

$\lim_{k\to\infty}\prod_{l=0}^{k-1}P(k-l) = \mathbf{1}_R c$, *where constant vector* $c = [c_1, 0, \cdots, 0] \in \Re^{1\times R}$ *with* $c_1 \in \Re^{1\times r_1}$.

Lemma 2.3. *(Qu et al., 2004b) Given sequences of row stochastic matrices* $P(k) \in \Re^{R\times R}$ *and* $P'(k) \in \Re^{R\times R}$*, where* $P(k)$ *is in the lower-triangular structure and* $P'(k)$ *satisfying* $P'_{ii}(k) \geq \varepsilon_p > 0$*. Then,*

$$\lim_{k\to\infty}\prod_{l=0}^{k-1}P(k-l)P'(k-l) = \mathbf{1}_R c_1,$$

if and only if $\lim_{k\to\infty}\prod_{l=0}^{k-1}P(k-l) = \mathbf{1}_R c_2$*, where* c_1 *and* c_2 *are constant vectors.*

3. Problem Formulation

Consider a group of MIMO dynamical systems given by

$$\dot{x}_i = A_i x_i + B_i u_i, \quad y_i = C_i x_i, \quad \dot{\eta}_i = g_i(\eta_i, x_i), \tag{1}$$

where $i = 1, \cdots, q$, $l_i \geq 1$ is an integer, $x_i \in \Re^{l_i m}$, $\eta_i \in \Re^{n_i - l_i m}$, J_k is the kth order Jordan canonical form given by

$$J_k = \begin{bmatrix} -1 & 1 & 0 & \cdots & 0 & 0 \\ 0 & -1 & 1 & \ddots & 0 & 0 \\ \vdots & \ddots & \ddots & \ddots & \ddots & \vdots \\ 0 & 0 & \cdots & -1 & 1 & 0 \\ 0 & 0 & 0 & \cdots & -1 & 1 \\ 0 & 0 & 0 & \cdots & 0 & -1 \end{bmatrix} \in \Re^{k\times k},$$

$A_i = J_{l_i} \otimes I_m \in \Re^{(l_i m)\times(l_i m)}$, $B_i = \begin{bmatrix} 0 \\ I_m \end{bmatrix} \in \Re^{(l_i m)\times m}$, $C_i = \begin{bmatrix} I_m & 0 \end{bmatrix} \in \Re^{m\times(l_i m)}$, u_i is the cooperative control law to be designed, and subsystem $\dot{\eta}_i = g_i(\eta_i, x_i)$ is input-to-state stable. Without loss of any generality, in this paper we consider the case that $l_1 = l_2 = \cdots = l_q = l$.

Remark 3.1. *Model (1) defines a general class of MIMO dynamical systems with finite but arbitrary relative degree. The system models considered in most of recent works (Jadbabaie et al., 2003)(Lin et al., 2004)(Saber and Murray, 2003) are either single or double integrator, which can be transformed into the given canonical model (1) with relative degree 1 or 2, respectively. In general, for a robot whose dynamics are given by* $\dot{\phi}_i = f_i(\phi_i, v_i)$, $y_i = h_i(\phi_i)$*, it is possible to find a diffeomorphic state transformation* $x_i = T_i(\phi_i)$ *and a decentralized control law* $v_i(\phi_i, u_i)$ *such that its dynamics are transformed into (1). It is straightforward to verify that an input-output feedback linearizable* *system with stable internal dynamics can be transformed into (1). For non-holonomic mobile robot, it is also possible to transform its model into (1) by using dynamic feedback linearization technique under some mild conditions.* ◇

The problem of making the systems' outputs and the states of individual systems converge to the same steady state has been studied, such as in (Jadbabaie et al., 2003)(Lin et al., 2004) for single integrator models and in (Qu et al., 2004a)(Qu et al., 2004b) for general dynamical systems like (1). In this paper, the objective is to synthesize a general class of cooperative controls and find the less-restrictive conditions on the connectivity requirements such that systems' different channels (outputs) as well as the corresponding states converge to different steady states. For ease of reference, we call this as multi-objective convergence.

4. Multi-Objective Cooperative Control

In general, we assume that the system sensors have a limited field of view, such as cone like of view, in that the system can only get the information from other systems in a relative direction and distance of itself. The control u_i for the ith system will be determined according to the output feedback information of itself as well as that of its near neighbors. The connectivity topology among individual systems can be represented by a signal transmission matrix, $S(t) = [S_{ij}(t)] \in \Re^{q \times q}$, where $S_{ii} = 1$ which means that system always has the information from itself; $S_{ij} = 0$ if no signal is received by the ith system from the jth system, otherwise $S_{ij} = 1$. If the directed sensor graph is strongly connected, that is, the signal transmission matrix $S(t)$ is irreducible, the convergence result has been obtained in (Lin et al., 2004) for linear system with single integrator model, and it is further extended to a general class of systems (1) (Qu et al., 2004a). However, this connectivity requirement is still too strong for cooperative control of a group of systems considering the limitations of sensor resources and communication burdens. This motivates us to study whether the overall system will still converge even when the directed sensor graphs are always not strongly connected (that is, the signal transmission matrices are reducible). When the control objective is to make all the states of individual systems converge to the same steady state, the question has been addressed in (Qu et al., 2004b) and a less-restrictive condition on the irreducibility of signal transmission matrix was given. In this section, we will study the multi-objective convergence problem and further extend the work in (Qu et al., 2004b)(Qu et al., 2004a) by establishing less-restrictive conditions on connectivity requirements and proposing a criterion for the design of cooperative feedback control matrices.

4.1 The Proposed Cooperative Control

Let the cooperative control be given by the following linear equation: for $i = 1, \cdots, q$,

$$u_i = \sum_{j \in N_i(t)} G'_{ij}(t) y_j = G_i(t) y, \tag{2}$$

where $N_i(t)$ denotes the set of labels of robot i's neighbor robots at time t, $G'_{ij} \in \Re^{m \times m}$ is the feedback gain matrix, $G'_{ij}(t) \geq 0$ and $G'_{ij} \neq 0$, $G_i = \left[\begin{array}{ccc} G_{i1} & \cdots & G_{iq} \end{array} \right]$ with $G_{ij} \in \Re^{m \times m}$ is the interconnection matrix satisfying the properties that $G_i \mathbf{1}_{mq} = \mathbf{1}_m$, and $y = [y_1^T \ \cdots \ y_q^T]^T$. It is easy to see from (2) that $G_{ij} = G'_{ij}$ if $j \in N_i(t)$, otherwise $G_{ij} = 0$. Assume that $G(t)$ will change over time according to physical surroundings, and $G_i(t)$ be piecewise constant for all i.

Let $\{t_k^G : k = 0, 1, \cdots\}$ with $t_0^G = t_0$ be the sequence of time instants at which $G(t)$ changes. That is, $G(t) = G(t_k^G)$ over the time interval $t \in [t_k^G, t_{k+1}^G)$. If there are only finite changes for $G(t)$, that is, for $t > t_i^G$, $G(t) = G(t_i^G)$, we can always partition the remaining time to generate an infinite time interval $[t_k^G, t_{k+1}^G)$. Suppose that $0 < t_{min} \leq t_{k+1}^G - t_k^G \leq t_{max}$. It follows from (1) and (2) that

$$\dot{x} = (A + D(t))x, \tag{3}$$

where $x = [x_1^T, \ \cdots, \ x_q^T]^T \in \Re^{N_q}$, $N_q = mql$, $x_i = [x_{i1}^T, x_{i2}^T, \cdots, x_{il}^T]^T \in \Re^{ml}$, $x_{ij} = [x_{ij1}, x_{ij2}, \cdots, x_{ijm}]^T \in \Re^m$ with $i = 1, \cdots, q$, and $j = 1, \cdots, l$, $A = diag\{A_1, \ \cdots, A_q\} \in \Re^{N_q \times N_q}$, $C = diag\{C_1, \ \cdots, \ C_q\} \in \Re^{(mq) \times N_q}$, $B = diag\{B_1, \ \cdots, \ B_q\} \in \Re^{N_q \times (mq)}$, $G = \left[\begin{array}{ccc} G_1^T & \cdots & G_q^T \end{array} \right]^T \in \Re^{(mq) \times (mq}$, and $D = BGC$.

It follows from the special structures of matrices A and $D(t)$ that $A + D(t) = -I_{N_q} + \bar{D}(t)$, where $\bar{D}(t) = [\bar{D}_{ij}(t)]$ with

$$\bar{D}_{ii} = \begin{bmatrix} 0 & I_{(l-1)} \otimes I_m \\ G_{ii} & 0 \end{bmatrix} \in \Re^{lm \times lm}, \quad \bar{D}_{ij} = \begin{bmatrix} 0 & 0 \\ G_{ij} & 0 \end{bmatrix} \in \Re^{lm \times lm},$$

where $i, j = 1, \cdots, q, \ \ i \neq j$. To achieve the multi-objective convergence, we need the following assumption:

Assumption 4.1. *Assume that the feedback matrix $G(t)$ satisfying the condition that sub-blocks $G_{ij}(t) = diag\{G_{ij,ss}(t)\}, s = 1, \cdots, m$ for all i and j.*

Let us rearrange the states of overall system and lump the different output and its corresponding relative states together by defining a state transformation $z = Tx$ as follows: $z = [z_1^T, \cdots, z_m^T]^T$, where $z_i = [z_{i1}^T, z_{i2}^T, \cdots, z_{iq}^T]^T \in \Re^{ql}$, $z_{ij} =$

$[z_{ij1}, \cdots, z_{ijl_j}]^T \in \Re^l$, $i = 1, \cdots, m$, $j = 1, \cdots, q$, and

$$\begin{array}{llllllll} z_{i1}: & z_{i11} = x_{11i}, & & z_{ij}: & z_{ij1} = x_{j1i}, & & z_{iq}: & z_{iq1} = x_{q1i} \\ & z_{i12} = x_{12i} & & & z_{ij2} = x_{j2i}, & & & z_{iq2} = x_{q2i}, \\ & \vdots & \cdots & & \vdots & \cdots & & \vdots \\ & z_{i1l} = x_{1li} & & & z_{ijl} = x_{jli}, & & & z_{iql} = x_{qli}. \end{array} \tag{4}$$

It is easy to see that T is a permutation matrix, which permutes the rows of x to obtain z. It follows from (3) and (4) that

$$\dot{z} = -(I_{N_q} - \tilde{D}(t))z, \tag{5}$$

where $\tilde{D} \stackrel{\triangle}{=} T\bar{D}T^{-1} = diag\{\tilde{D}_{11}, \cdots, \tilde{D}_{mm}\}$ with $(s = 1, \cdots, m)$

$$\tilde{D}_{ss} = \begin{bmatrix} 0 & I_{l-1} & 0 & 0 & \cdots & 0 & 0 \\ G_{11,ss} & 0 & G_{12,ss} & 0 & \cdots & G_{1q,ss} & 0 \\ 0 & 0 & 0 & I_{l-1} & \cdots & 0 & 0 \\ G_{21,ss} & 0 & G_{22,ss} & 0 & \cdots & G_{2q,ss} & 0 \\ \vdots & \vdots & \vdots & \vdots & \vdots & \vdots & \vdots \\ 0 & 0 & 0 & 0 & \cdots & 0 & I_{l-1} \\ G_{q1,ss} & 0 & G_{q2,ss} & 0 & \cdots & G_{qq,ss} & 0 \end{bmatrix} \in \Re^{ql \times ql}.$$

4.2 Conditions on Multi-Objective Convergence

In this subsection, we establish a set of less-restrictive conditions on feedback matrix $G(t)$ such that the multi-objective convergence is achieved when the signal transmission matrices are reducible. It is shown in (Minc, 1988) that, if $S(t)$ is reducible, then there is a permutation matrix $T_1 \in \Re^{q \times q}$ such that $S_{T_1}(t) = T_1^T S(t) T_1$ is in the lower-triangular structure, that is

$$S_{T_1}(t) = \begin{bmatrix} S_{T_1,11}(t) & 0 & \cdots & 0 \\ S_{T_1,21}(t) & S_{T_1,22}(t) & \cdots & 0 \\ \vdots & \vdots & \ddots & \vdots \\ S_{T_1,k1}(t) & S_{T_1,k2}(t) & \cdots & S_{T_1,kk}(t) \end{bmatrix},$$

where $S_{T_1,ii} \in \Re^{q_i \times q_i}$, $\sum_{i=1}^{k} q_i = q$, and $S_{T_1,ii}(t)$ are irreducible. Correspondingly, we have augmented permutation matrices $T_2 = T_1 \otimes I_m \in \Re^{qm \times qm}$ and $T_3 = \text{diag}\{T_1 \otimes I_l, \cdots, T_1 \otimes I_l\} \in \Re^{qlm \times qlm}$, such that $G_{T_2}(t) = T_2^T G(t) T_2$ is also in the lower-triangular structure, and the state transformation is $z = T_3 \xi$. To this end, the system dynamics (5) become

$$\dot{\xi} = -(I_{N_q} - T_3^T \tilde{D} T_3)\xi. \tag{6}$$

Remark 4.1. *Generally, the case of $q_i \neq 1$ means that the q systems reformulate k subgroups with $k < q$.* ◇

Example 4.1. *Consider the case of 4 robots with $m = 2$ and $l = 1$. We have $N_q = 8$. For states $x = [x_1^T, x_2^T, x_3^T, x_4^T]^T$ with $x_i = [x_{i1}, x_{i2}]^T$, $i = 1, \cdots, 4$, it is easy to find transformation T such that $z = Tx = [z_1^T, z_2^T]^T$ with $z_1 = [x_{11}, x_{21}, x_{31}, x_{41}]^T$ and $z_1 = [x_{12}, x_{22}, x_{32}, x_{42}]^T$. Thus, we have $\tilde{D} = T\bar{D}T^{-1} = diag\{\tilde{D}_{11}, \tilde{D}_{22}\}$, with $\tilde{D}_{ss} = [G_{ij,ss}]$, $s = 1, 2$, $i, j = 1, \cdots, 4$. Let the reducible signal transmission matrix S and permutation matrix T_1 be*

$$S = \begin{bmatrix} 1 & 1 & 0 & 1 \\ 0 & 1 & 0 & 0 \\ 1 & 0 & 1 & 0 \\ 0 & 0 & 1 & 1 \end{bmatrix}, \quad T_1 = \begin{bmatrix} 0 & 1 & 0 & 0 \\ 1 & 0 & 0 & 0 \\ 0 & 0 & 1 & 0 \\ 0 & 0 & 0 & 1 \end{bmatrix}.$$

Then we have $S_{T_1} = \begin{bmatrix} S_{T_1,11} & 0 \\ S_{T_1,21} & S_{T_1,22} \end{bmatrix}$, with

$$S_{T_1,11} = 1, \quad S_{T_1,21} = \begin{bmatrix} 1 \\ 0 \\ 0 \end{bmatrix}, \quad S_{T_1,22} = \begin{bmatrix} 1 & 0 & 1 \\ 1 & 1 & 0 \\ 0 & 1 & 1 \end{bmatrix},$$

where $S_{T_1,11}$ and $S_{T_1,22}$ are irreducible. To this end, four robots is reformulated into two subgroups, that is, group 1 = $\{2\}$ and group 2 = $\{1, 3, 4\}$. We have the corresponding $T_2 = T_1 \otimes I_2$ such that $G_{T_2} = \begin{bmatrix} G_{T_2,11} & 0 \\ G_{T_2,21} & G_{T_2,22} \end{bmatrix}$, with $G_{T_2,11} = G_{22}$ and

$$G_{T_2,21} = \begin{bmatrix} G_{12} \\ 0 \\ 0 \end{bmatrix}, \quad G_{T_2,22} = \begin{bmatrix} G_{11} & 0 & G_{14} \\ G_{31} & G_{33} & 0 \\ 0 & G_{43} & G_{44} \end{bmatrix}.$$

Also, $T_3 = diag\{T_1, T_1\}$, and $\tilde{D}_{T_3} = T_3^T \tilde{D} T_3 = \begin{bmatrix} T_1^T \tilde{D}_{11} T_1 & 0 \\ 0 & T_1^T \tilde{D}_{22} T_1 \end{bmatrix}$, where $T_1^T \tilde{D}_{ss} T_1 = \begin{bmatrix} \tilde{D}_{sT_3,11} & 0 \\ \tilde{D}_{sT_3,21} & \tilde{D}_{sT_3,22} \end{bmatrix}$, with $\tilde{D}_{sT_3,11} = G_{22,ss}$ and

$$\tilde{D}_{sT_3,21} = \begin{bmatrix} G_{12,ss} \\ 0 \\ 0 \end{bmatrix}, \quad \tilde{D}_{sT_3,22} = \begin{bmatrix} G_{11,ss} & 0 & G_{14,ss} \\ G_{31,ss} & G_{33,ss} & 0 \\ 0 & G_{43,ss} & G_{44,ss} \end{bmatrix}.$$

◇

In what follows, we present the main result of the paper and give a less-restrictive condition on the design of cooperative control feedback matrix.

Assumption 4.2. *Assume that the signal transmission matrix $S(t_k^G)$ is reducible for almost all k, and $G_{T_2}(t_k^G)$ is in the lower-triangular form. Define a subsequence $\{s_v, v = 0, 1, \cdots, \infty\}$ of $\{0, 1, 2, \cdots, \infty\}$, $s_0 = 0$ and $\lim_{v\to\infty} s_v = +\infty$, such that for all time instants $t_{s_v}^G$ permutation matrices $T_2(t_{s_v}^G)$ take the same value. Assume that $G_{T_2}(t_{s_v}^G)$ has the following properties: (i) for the nonzero element in the i_1th row i_2th column of irreducible matrix $S_{T_1,ii}(t_{s_v}^G)$, the corresponding i_1th row i_2th column sub-blocks of matrix $G_{T_2,ii}(t_{s_v}^G)$ which is $m \times m$ dimensional has all its diagonal elements positive; (ii) for every $i > 1$, there is at least one j such that $G_{T_2,ij}(t_{s_v}^G) \succ 0$, $j < i$.*

Theorem 4.1. *Consider the cooperative control of system (1) using (2) under assumptions 4.1 and 4.2. Then,*

$$\lim_{t\to\infty} x(t) = \mathbf{1}_{ql} \otimes cx(0), \tag{7}$$

$c \in \Re^{m\times N_q}$ is a constant matrix.

Proof: For briefness, we denote $f(t_k^G) \triangleq f(k)$ for symbol f. The proof of (7) is equivalent to that of

$$\lim_{t\to\infty} z(t) = \text{diag}\{\mathbf{1}_{ql}c_1, \cdots, \mathbf{1}_{ql}c_m\}z(0), \tag{8}$$

where $c_1, \cdots, c_m \in \Re^{1\times ql}$ are row vectors and $c = \text{diag}\{c_1, c_2, \cdots, c_m\} \in \Re^{m\times N_q}$. It follows that the solution of (6) is given by

$$\xi(t_{k+1}^G) = \prod_{l=0}^{k} P(k-l)\xi(t_0^G), \tag{9}$$

where $P(i) = e^{-(I-T_3^T(i)\tilde{D}(t_i^G)T_3(i))(t_{i+1}^G - t_i^G)}$, $i = 0, \cdots, k$. It follows from (9) and $z(k) = T_3(k)\xi(k)$ that

$$z(t_{k+1}^G) = \prod_{l=0}^{k} T_3(k-l)P(k-l)T_3(k-l)^T z(t_0^G), \tag{10}$$

Since $T_3(k) = \text{diag}\{T_1, \cdots, T_1\}$ and $\tilde{D}(k)$ are in the diagonal structure, we have $P_{ss}(i) = e^{-(I-T_1^T(i)\tilde{D}_{ss}(i)T_1(i))(t_{i+1}^G - t_i^G)}$, and

$$z_s(t_{k+1}^G) = \prod_{l=0}^{k} T_1(k-l)P_{ss}(k-l)T_1(k-l)^T z_s(t_0^G), \tag{11}$$

where $s = 1, \cdots, m$. Thus, we only need to show that $\lim_{t\to\infty} z_s(t) = \mathbf{1}_{ql}c_s z_s(0)$. It suffices to prove that

$$\lim_{k\to\infty} \prod_{l=0}^{k} T_1(k-l)P_{ss}(k-l)T_1(k-l)^T = \mathbf{1}_{ql}c_s. \tag{12}$$

It follows from $G_{T_2}(t_k^G)$ is in the lower-triangular structure that $\tilde{D}_{sT_3} = T_1^T(k)\tilde{D}_{ss}(t_k^G)T_1(k)$ and $P_{ss}(k)$ are also in the lower-triangular structure. Moreover, $P_{ss}(k)$ is row-stochastic matrix and its diagonal elements are lower-bounded by a positive value (Freedman, 1983). By assumption 4.2 that the i_1th row i_2th column sub-blocks $G_{T_2,ii}(t_{s_v}^G)$ has all its diagonal elements positive and $S_{T_1,ii}(t_{s_v}^G)$ is irreducible, we have that $\tilde{D}_{sT_3,ii}(t_{s_v}^G)$ is irreducible by lemma 2.1 and $P_{ss,ii}(s_v) > 0$ (Qu et al., 2004a). On the other hand, $G_{T_2,ij}(t_{s_v}^G) \succ 0$ leads to $\tilde{D}_{sT_3,ij}(t_{s_v}^G) \succ 0$ and $P_{ss,ij}(s_v) \succ 0$. It then follows from assumption 4.2 and lemma 2.2 that

$$\lim_{v\to\infty} P_{ss}(s_v)P_{ss}(s_{v-1})\cdots P_{ss}(s_0) = \mathbf{1}_{ql}c_s', \tag{13}$$

where c_s' is a constant vector. Define $P_{ss}'(s_v) = T_1(s_v)^T T_1(s_v - 1)P_{ss}(s_v - 1)T_1^T(s_v - 1)\cdots T_1(s_{v-1}+1)P_{ss}(s_{v-1}+1)T_1^T(s_{v-1}+1)T_1(s_{v-1})$. It then follows from (13) and the fact that $P_{ss}'(s_v)$ has positive diagonal elements that, (12) can be proved using lemma 2.3. This completes the proof. □

5. An Illustrative Example

Consider a group of three nonholonomic 4-wheel differential driven mobile robots. By taking the robot "hand" position as the guide point, whose model can be feedback linearized to a double integrator with a stable internal dynamics (Qu et al., 2004a):

$$\dot{z}_{i1} = z_{i2}, \quad \dot{z}_{i2} = v_{i2},$$

where $i = 1,2,3$, $z_{i1} = [z_{i11}, z_{i12}]^T \in \Re^2$ is the position of particle i, $z_{i2} = [z_{i21}, z_{i22}]^T \in \Re^2$ its velocity, and $v_i = [v_{i1}, v_{i2}]^T \in \Re^2$ its acceleration inputs. The cooperative control objective is that all particles move to the same position but with different horizontal and vertical coordinates value. Define the state and input transformations $x_{i1} = z_{i1}$, $x_{i2} = x_{i1} + z_{i2}$, $v_i = -2x_{i2} + x_{i1} + u_i$. Then system model can be transformed into $\dot{x}_{i1} = -x_{i1} + x_{i2}$, $\dot{x}_{i2} = -x_{i2} + u_i$,where $u = Gy$ with $y = [x_{11}^T, x_{21}^T, x_{31}^T]^T$. To this end, the cooperative control method in this paper can be used for the design of G. For illustration purpose, assume that two kinds of leader-follower situations appear during the robot motion process: (i) robot 1 as the leader, robot 2 follows robot 1 and robot 3 follows robot 2; (ii) robot 2 as the leader, robot 1 follows robot 2 and robot 3 follows robot 1. After permutation T_1 and T_2, the corresponding cooperative feedback matrices take the form

$$G_{T_1} = \begin{bmatrix} G_{11} & 0 & 0 \\ G_{21} & G_{22} & 0 \\ 0 & G_{32} & G_{33} \end{bmatrix}, \quad G_{T_2} = \begin{bmatrix} G_{22} & 0 & 0 \\ G_{12} & G_{11} & 0 \\ 0 & G_{31} & G_{33} \end{bmatrix}.$$

To satisfy the condition in theorem 4.1, let G_{T_1} and G_{T_2} be chosen as

$$\begin{bmatrix} 1 & 0 & 0 & 0 & 0 & 0 \\ 0 & 1 & 0 & 0 & 0 & 0 \\ 0.2 & 0 & 0.8 & 0 & 0 & 0 \\ 0 & 0.2 & 0 & 0.8 & 0 & 0 \\ 0.2 & 0 & 0 & 0 & 0.8 & 0 \\ 0 & 0.2 & 0 & 0 & 0 & 0.8 \end{bmatrix}.$$

The initial positions are $[6,3]^T$, $[2,5]^T$ and $[4,2]^T$, respectively. Figure 1 shows the convergence of robots' position, which verifies the proposed design in this paper.

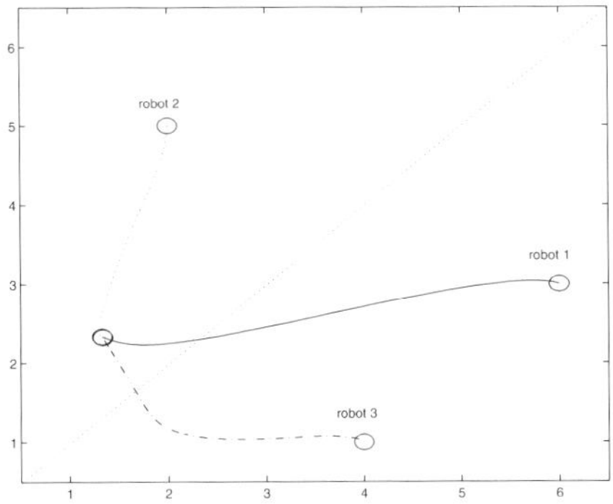

Figure 1. Convergence of positions under cooperative control

6. Conclusion

In this paper, the multi-objective convergence problem has been studied for a general MIMO dynamical systems. The obtained less-restrictive conditions on the design of cooperative feedback matrices require neither the strong connectivity of system sensor graphs nor the fixed communication structure for robots. The proposed method can be applied to the formation stabilization and formation tracking control of robots.

References

Freedman, D. (1983). *Markov Chains*. Springer-Verlag, New York.

Jadbabaie, A., Lin, J., and Morse, A. S. (2003). Coordination of groups of mobile autonomous agents using nearest neighbor rules. *IEEE Trans. on Automatic Control*, 48:988–1001.

Lin, Z., Brouchke, M., and Francis, B. (2004). Local control strategies for groups of mobile autonomous agents. *IEEE Trans. on Automatic Control*, 49:622–629.

Minc, H. (1988). *Nonnegative Matrices*. John Wiley & Sons, New York, NY.

Moreau, L. (2003). Leaderless coordination via bidirectional and unidirectional time-dependent communication. In *Proceedings of the 42nd IEEE Conference on Decision and Control*, Maui, Hawaii.

Qu, Z., Wang, J., and Hull, R. A. (2004a). Cooperative control of dynamical systems with application to mobile robot formation. In *The 10th IFAC/IFORS/IMACSIFIP Symposium on Large Scale Systems: Theory and Applications*, Japan.

Qu, Z., Wang, J., and Hull, R. A. (2004b). Products of row stochastic matrices and their applications to cooperative control for autonomous mobile robots. In *Technique report; also submitted to 2005 American Control Conference*.

Ren, W. and Beard, R. W. (2004). Consensus of information under dynamically changing interaction topologies. In *Proceedings of the American Control Conference*, Boston.

Reynolds, C. W. (1987). Flocks, herds, and schools: a distributed behavioral model. *Computer Graphics (ACM SIGGRAPH 87 Conference Proceedings)*, 21(4):25–34.

Saber, R. O. and Murray, R. M. (2003). Consensus protocols for networks of dynamic agents. In *Proceedings of the American Control Conference*, Denver, CO.

Vicsek, T., Czirok, A., Jacob, E. B., Cohen, I., and Shochet, O. (1995). Novel type of phase transition in a system of self-driven particles. *Physical Review Letters*, 75:1226–1229.

Wolfowitz, J. (1963). Products of indecoposable, aperiodic, stochastic matrices. *Proc. Amer. Mathematical Soc.*, 14:733–737.

LEVELS OF MULTI-ROBOT COORDINATION FOR DYNAMIC ENVIRONMENTS

Colin P. McMillen, Paul E. Rybski, Manuela M. Veloso
School of Computer Science
Carnegie Mellon University
Pittsburgh, PA 15213, U.S.A.
mcmillen@cs.cmu.edu, prybski@cs.cmu.edu, veloso@cs.cmu.edu

Abstract RoboCup, the international robot soccer competition, poses a set of extremely difficult challenges for multi-robot systems. To be competitive in RoboCup's rapidly-changing, dynamic, adversarial environment, teams need to make use of effective coordination strategies. We describe some of our experiences with effective coordination of robots teams and introduce several levels of strategies which encapsulate coordination from the level of individual robots to synchronized coordination of the entire team.

Keywords: Adversarial environments, robot soccer, multi-robot coordination

1. Introduction

The RoboCup robot soccer competition is a domain in which teams must address the challenges of real-time perception, cognition, and action. Robots must be able to operate in a very dynamic environment in which they must reason not only about the actions of their teammates, but also about the actions of a team of adversarial agents. Teams of robots that operates without an effective teamwork strategy are likely to hamper each other's efforts and perform as an inferior team. Our primary research interests are in exploring the scientific challenges of developing effective teamwork strategies for autonomous robotic systems where all sensing and cognition is done on-board the robot.

We have had extensive experience with robot positioning ranging from early work in simulated robot soccer players (Veloso et al., 1999) to the Sony AIBO league (Uther et al., 2002). In this previous work, we made use of artificial potential field methods and have found them to be a very powerful way of representing multiple constraints when positioning robots. However, there are a number of limitations in the kinds of behaviors that potential fields can express. We are actively exploring other coordination strategies which we will describe in more detail.

L.E. Parker et al. (eds.),
Multi-Robot Systems. From Swarms to Intelligent Automata. Volume III, 53–64.

This paper focuses on developing team strategies for robots that compete in the RoboCup (Kitano et al., 1997) legged league. We discuss various approaches for teamwork and cooperation, and describe some empirical results from experiments.

1.1 Related Work

Potential field methods have been used very successfully for navigation tasks such as obstacle avoidance (Khatib, 1985). This idea has been extended such that a group of robots can maintain formations while using only local information in their potential calculations (Balch and Arkin, 1998, Balch and Hybinette, 2000).

Several groups have encoded domain-specific heuristics into potential fields. In the RoboCup domain, potential fields can be constructed that guide robots to an area near the opponent's goal or to an open position that is well-suited for pass reception (Castelpietra et al., 2001, Veloso et al., 1999, Weigel et al., 2001). A behavior architecture that relies on potential fields for motion planning and action selection is described in (Laue and Röfer, 2004). Their approach has been applied to real robots, including the German Team of the RoboCup legged league.

Potential field methods have several limitations that have been reported in the literature, including susceptibility to local minima and oscillations (Koren and Borenstein, 1991). An approach known as forward chaining dynamically reshapes the potential field using heuristics that guide the robot to the goal through a series of subgoals or waypoints that attempt to avoid local minima (Bell and Weir, 2004). The forward chaining approach is particularly interesting to us because it utilizes the idea of a positioning function that changes over time.

1.2 Coordination in Robot Soccer

We have been researching coordination strategies for the different RoboCup robot soccer leagues for several years. RoboCup robot soccer in general (Kitano et al., 1997) offers a very challenging research environment for multi-robot systems. One of the most interesting aspects of the RoboCup leagues is that we change the rules of the game and the playing environment every year in order to increase the difficulty of the problem towards matching real setups as much as possible.

Of particular interest to this paper is our work in the RoboCup legged league with the Sony AIBO four-legged robots, which need to be fully autonomous with on-board sensing, computation, and action. The legged league in particular has gone through several changes since 1998. Some of the most significant changes occurred in 2002 and have made the most impact in our multi-robot research efforts.

Communicating robots: The AIBO robots are now equipped with the ability to communicate wirelessly. In the initial years, when the robots could not communicate, we achieved teamwork through vision of the position of the ball – the robots' behaviors were vision-servoed. Three attacker robots searched for the ball; as soon as they saw the ball, they would move towards it and then move the ball towards the opponent goal. Because the ball was often not in the robots' field of view, due to the occlusion by other robots and the robots' own search for localization markers, not all the robots could see the ball. Teamwork was a property that emerged due to this discontinuity of ball perception: when one robot saw the ball, it would move toward the ball, tending to block the ball from the view of its teammates. Because of this occlusion, the teammates would remain spread out from the "attacker." Now that communication is available between robots, we have researched methods of sharing information (Roth et al., 2003) and dynamically changing robot roles and positioning (Vail and Veloso, 2003). In this paper, we discuss the different levels of dynamic coordination necessary in the presence of skilled opponent teams. We present the solutions that we have developed and plan to continue researching.

World space increase: The field's size has increased by approximately 50% from its initial size, and the number of robots in a team has increased from three to four. The increase in field size makes it infeasible for robots to see across the entire field. In the initial smaller field, individual robots could recognize objects across the complete field. Modeling the world state is now a task that needs to combine a robot's own perception and the communicated information from other robots. Multiple robots need to build an accurate world model and select joint actions that fit a team policy.

Rules of the game: The rules of the game set constraints on the legal positioning and actions of robots. For example, only one robot is allowed to defend the goal area. This type of rule creates hard constraints on the dynamic positioning of team members. In addition, robots encounter difficult motion situations when surrounded by opponent robots. In these situations, a team member may need the help of robot teammates. This is a challenge that requires an effective dynamic coordination algorithm that monitors the progress of teammates. In addition, teamwork should change as a function of the opponent team, the specific state of the field, and the remaining time of the game.

The RoboCup legged league continues to motivate our research in multi-robot domains, inspiring incremental algorithmic successes and providing many issues to be addressed. Interestingly, the more we work on this adversarial multi-robot coordination problem, the more we understand how the problems we face go well beyond robot soccer and are of relevance to multi-robot systems in complex environments. In this paper, we present our findings aiming at such an abstract level.

2. Dynamic Multi-Robot Coordination

Over the past few years, teams have experimented with different methods of team coordination. Many of these strategies involve keeping teammates from running into each other and placing teammates in good locations on the field so that they can be in good positions to receive passes or go after a free ball. While there have been some good approaches, no one strategy has emerged as being clearly superior to all others. One reason for this is that several different coordination strategies are likely to be applicable in a single situation. Since some strategies may work better than others, a team that selects the superior strategy will be at an advantage. Thus, one of the most important problems to address when designing a multi-robot soccer team is selecting the kind of coordination strategy that will be used during the game. Teams may choose to use a fixed coordination strategy defined *a priori*, but if chosen poorly, a fixed strategy may not fare well against the strategy of the other team. Thus, an important extension to the research problem of coordination strategies is the ability for a team to dynamically change their strategy at runtime to adapt to their opponents' strengths and weaknesses.

Dynamically selecting a different strategy depending on the situation can be very powerful technique, but can be very challenging to implement well. Robots that use a dynamic coordination system must be able to perceive and properly evaluate the state of the world as well as the state of their own progress. This information is vital when making the decision to switch from a poorly performing strategy to one that could potentially work better.

We have identified several different levels for dynamic coordination that can be applied to a robotic team. These include:

- A "first-order" approach, where the robots use a fixed coordination strategy and each robot modifies the parameters of its behavior according to the world state.

- A "second-order" approach, where the robots have multiple ways of handling different situations. In order to utilize a second-order strategy, the robots must be able to evaluate the world state so that they can choose between the different behaviors they have at their disposal.

- A "third-order" approach, where the robots have several different team strategies, or "plays," which describe the coordinated actions of all of the robots together. Depending on the world state, different plays may apply; the team collectively decides upon the right behavior to apply in a given situation.

We have implemented methods for first- and second-order coordination strategies, a description of which is provided below. Currently, the third level of coordination has been implemented in our small-sized league (Bowling et al., 2004) but not yet on the AIBOs.

2.1 Changing Single Robot Parameters

We define the first-order coordination strategy as the ability for the robots to set their own behavior based on the state of the world. In this kind of system, each robot is programmed with a single behavior set which is used to control the robot's behavior in its environment.

We have tried two different methods for representing first-order coordination strategies. The first is a potential fields approach and the other is an approach that we call constraint-based positioning. In previous work (Vail and Veloso, 2003), we give a detailed description of our implementation of potential field-based coordination. In this approach, we use potential fields both to determine the role that each robot plays (attacker, supporting attacker, and defender) and also to determine where the supporting robots position themselves on the field of play. On efficiency issue with potential fields occurs when they are used to coordinate the actions of a team of robots in a very dynamic world. In this situation, the fields may need to be recomputed for each every new sensor reading. This does not tend to be true for implementations of potential fields that are used for navigation in more static environments. In general, however, it's possible for minor disturbances in the positions or strengths of individual attraction and repulsion fields to cause fairly significant changes in the local gradient surrounding the robot.

Constraint-based positioning is an approach to robot positioning that we have developed in the last year for the 2004 RoboCup competition. Under this regime, robots are still assigned roles using a potential function, but the field positions chosen by the supporting robots are subject to a set of constraints. This approach was developed because there are several hard constraints that we would like to enforce on the robots' positions which are difficult to specify clearly with potential fields. For instance, defender robots need to avoid their own goalie's defense box, because entering the defense box is a violation which will cause the robot to be removed from play for 30 seconds. Other constraints that we would like to enforce include not crossing in front of a robot that is about to take a shot on goal, not coming within a certain minimum distance of a teammate, and so on. Consider a situation in which a robot is near the defense zone and a teammate is directly approaching it. Should the robot move toward the goal, violating the defense-zone constraint, or stand still, violating the teammate-distance constraint? Our implementation of constraint-based positioning allows us to prioritize the constraints, so that the robot knows that entering the defense zone is a more serious violation than coming near a teammate. In theory, the priorities of these constraints could be represented as a potential field, but we have found that debugging the complex potential fields that result can be difficult. If no constraints are in danger of being violated, the robot can choose to move to a specific point that is chosen based on the current state of the world. In this case, the robot can still use potential fields to choose an open area on the field or to choose a path to navigate around local obstacles.

Our experience with RoboCup has been that a single positioning function defined for a particular role tends to be too limiting. Trying to capture all of the possible actions that a robot might accomplish can cause the complexity of the positioning function to grow beyond what is manageable. A soccer-playing robot might have multiple ways of approaching the goal, each of which has advantages depending on the relative position of the goalie and/or his other players. In some situations, the robot may want to try one approach and if it fails, try a different approach. Behaviors like these may be mutually exclusive and as such could be very difficult for a single function to capture.

2.2 Changing Single Robot Behaviors

An alternative is to factor the problem into subproblems and make multiple positioning functions available for the robot to use. In this case, a second-order decision process must exist whose purpose is to evaluate the state of the world and/or the current performance of the robot. This decision process is responsible for deciding which positioning function should be used in a particular situation.

Designing multiple behaviors such as these with potential fields requires that an entirely new set of potential attractor/repulsor nodes be defined for each of the new behaviors. A single set of nodes cannot be used for independent behaviors because the individual nodes are not independent of each other. They all affect one another.

Another challenge with potential fields is that in the case of multiple specific and possibly exclusive behavior sets, a robot may be expected to approach a very specific location on the field and stay there. Specifying a specific (x, y, θ) location on the field would be fairly straightforward for a constraint-based system to handle, but designing the potentials such that they push the robot to a specific location on the field can be a very challenging task. An extreme solution for the potential fields approach is to have a single potential attractor that pulls the robot to the specified point. This suggests that having control over the attraction/repulsion nodes and being able to turn them on and off as necessary would make the potential field approach work in this situation.

In a constraint-based system, the decision process evaluates the points on the field and chooses a specific location for the robot to reach. In both positioning methodologies, a higher-level decision process is in charge of selecting the specifics of the behavior set by evaluating the state of the environment and selecting the one with the highest chance of success.

3. Experimental Results

We have performed a set of experiments that show the need for second-order reasoning in the RoboCup domain. These experiments demonstrate that we can improve performance by having a higher-level decision process that changes the positioning strategy based on the environment. Specifically, we compare the performance of two positioning strategies under differing environmental conditions, and show that the strategy which is superior in one situation is inferior in the other situation.

In each experimental trial, we placed the ball in one of the front corners of the field, and two robots (on the same team) attempted to score a goal within thirty seconds. We chose this initial position of the ball because it has traditionally been difficult to score a goal from the front corner of the field. In this situation, it is not usually possible to score a goal by a single direct kick; trying to do so will often send the ball rolling into the opposite corner. From the other corner, the attacker may very well choose to execute another strong kick toward the goal, which can lead to a series of "ping-pong" kicks across the goal until the goalkeeper clears the ball or until noise in the environment causes the ball to roll into a different area of the field. The 30-second time limit only gives the robots enough time to execute approximately three to five kicks, so we feel that a goal scored within that time limit indicates that the robots were performing reasonably well during that trial.

In half of the trials, we placed a goalie robot in the defense zone, facing the corner where the ball was initially placed. The position chosen was the one that our own goalie would adopt if the ball were placed in that position. However, the goalie was paused, and therefore did not attempt to clear the ball or attempt to move from this initial position unless it was pushed by the other robots. In the other half of the trials, no goalie was placed on the field.

One of the two robots on the team (the attacker) was placed 75 cm away from the ball, facing the corner of the field. The supporting robot was positioned according to one of two different potential fields. Both fields simply contained a single linear attractor that pulled the supporter to a desired point. In the *side* potential field, the supporter was drawn toward a point on the opposite corner

(a)

(b)

Figure 1. Two of the four initial configurations used in the experimental trials. Image (a) shows the supporter in the center position with a stationary goalie present on the field. Image (b) shows the supporter in the side position with no goalie.

of the goal; in the *center* potential field, the supporter was drawn toward a center point about 100 cm from the front of the goal. See Figure 1 for pictures of the initial configurations of the field, including the supporter positioning induced by the two different potential fields.

We ran 40 trials for all four different possible setups (with or without goalie, combined with center or side positioning), for a total of 160 trials. For each trial, the success or failure of the run was recorded. If the run was a success (i.e., it terminated in a goal), we also recorded the amount of time it took for the robots to score the goal.

Each run started by unpausing the attacker robot; the 30-second timer was started as soon as the attacker touched the ball. If any robot crashed or ran out of batteries during a trial, the robot was rebooted and the trial was restarted from the beginning. Normal RoboCup penalties, such as player pushing, goalie pushing, and holding, were not enforced. If the ball was knocked out of the field, it was immediately placed back in-bounds at the place where it went out, as per the RoboCup 2004 rules.

The results of these experimental runs are summarized in Table 1. Figure 2 shows the individual completion times for every trial. Note that the results are only shown for the runs that were counted as successes; therefore, each graph has a different number of points plotted.

In the no-goalie case, the side positioning succeeded slightly more often than the center positioning, and the mean time per success was significantly lower for the side positioning. (Statistical significance of the mean time determined by Student's two-tailed t-test, with $p = 0.001$.) However, in the runs with the goalie, the center positioning significantly outperformed the side positioning, with a higher number of successes and a faster mean time per success.

	Successes	Failures	Mean Time per Success
Side positioning, no goalie	31	9	9.97s
Center positioning, no goalie	27	13	16.91s
Side positioning, with goalie	12	28	23.63s
Center positioning, with goalie	17	23	18.55s

Table 1. Summary of the results obtained in the experimental trials.

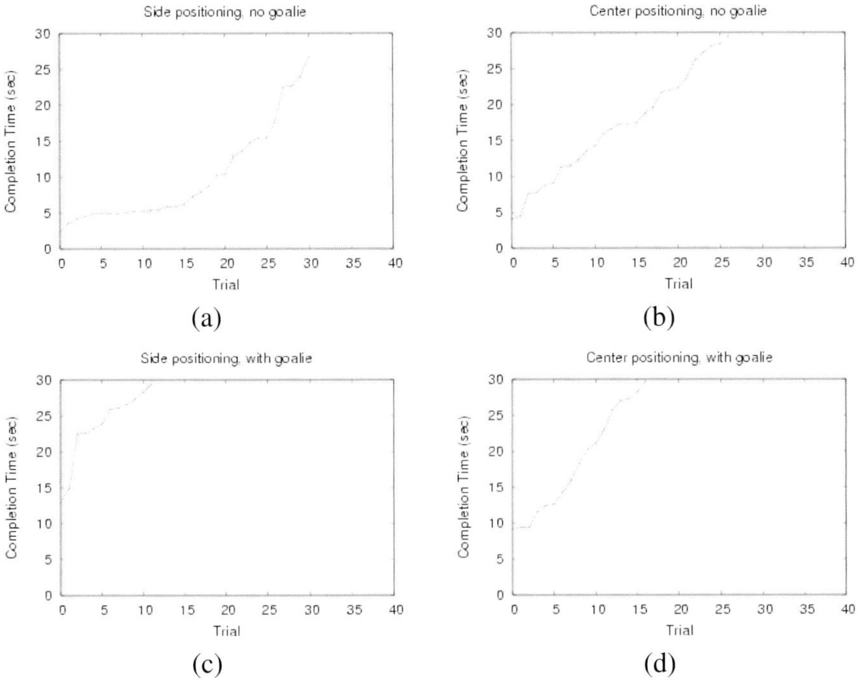

Figure 2. Graphs showing the amount of time it took to successfully score a goal. Each trial was stopped after 30 seconds if a goal had not yet been scored. Graphs (a) and (b) show the results for the no-goalie case; graphs (c) and (d) show the results for the with-goalie case. Trials are sorted from fastest to slowest completion time.

(Statistical significance of the mean time determined by Student's two-tailed t-test, with $p = 0.047$.)

The advantages and disadvantages of each position are easily explained through a qualitative analysis. The side position does much better in the no-goalie case because the position of the supporter puts it in a very good location to intercept the attacker's kick. After a successful interception, a single head kick is usually sufficient to score a quick goal. The center positioning does not enable the easy interception of a missed shot, so it is more likely that the ball will end up in the opposite corner and require more time before a goal can be scored.

However, when a goalie is added to the field, the weaknesses of the side positioning become apparent. The initial kick often bounces off the goalie and stops close to the center of the field, instead of traveling across the field to the other side. In this situation, the supporter positioned in the center is much more likely to be able to assist the attacker. Furthermore, it is difficult for the side-positioned supporter to react quickly to changes in the ball's location, since the supporter's view of the ball is often occluded by the goalie. The center positioning is a more general approach that allows the supporter to chase down the ball relatively quickly wherever it goes on the field, while the side positioning is superior in the special case where the opposing goalie is temporarily outside the defense box.

Though the center positioning is the approach that we would prefer the majority of the time, there is a definite benefit to being able to use side positioning to exploit the situation when the goalie is not guarding the goal. For example, one of the only two goals scored in the (very defensive) final game of the 2004 US Open occurred when the opposing goalie temporarily left the defense zone and was inadvertently blocked from returning to the goal by another robot that had gotten behind it. The results presented in this section suggest that there is definitely a benefit to be gained from using second-order reasoning in multi-robot systems, especially in an adversarial, dynamic environment.

4. Conclusion / Future Work

In this paper, we have proposed a classification scheme that identifies various levels of dynamic multi-robot coordination. We have provided examples showing the limitations of first-order coordination strategies in the robot soccer domain, and presented experimental results that show that there is a substantial benefit to our use of second-order reasoning about team coordination.

In the future, we intend to improve upon our existing coordination strategies by adding third-order functionality to our team. We plan to take inspiration from the idea of using a playbook for team coordination, which has been a successful strategy in the RoboCup small-size league (Bowling et al., 2004). The effectiveness of playbooks in the small-size league is largely due to the fact that this league makes use of an overhead camera and so the state of the entire team can be very easily determined. The legged league has no such overhead camera system and so a team state estimate must be computed in a distributed fashion by merging the local sensory information from each of the robots. We are actively researching methods for accomplishing this task so that we can pursue the development of third-order coordination strategies, such as a playbook, for our RoboCup legged league team.

Acknowledgments

The authors would like to thank the other team members of CMPack'04: Sonia Chernova (team leader), Douglas Vail, Juan Fasola, and Scott Lenser. The authors would also like to thank James Bruce for his assistance with the development of the team.

This work was supported by United States Department of the Interior under Grant No. NBCH-1040007. The content of the information in this publication does not necessarily reflect the position or policy of the Defense Advanced Research Projects Agency (DARPA), US Department of Interior, US Government, and no official endorsement should be inferred.

References

Balch, T. and Hybinette, M. (2000). Social potentials for scalable multirobot formations. In *Proceedings of the IEEE International Conference on Robotics and Automation*, volume 1, pages 73–80.

Balch, T. R. and Arkin, R. C. (1998). Behavior-based formation control for multiagent robot teams. *IEEE Transactions on Robotics and Automation*, 14(6):926–939.

Bell, G. and Weir, M. (2004). Forward chaining for robot and agent navigation using potential fields. In *Proceedings of the 27th conference on Australian computer science*, volume 26, pages 265–274, Dunedin, New Zealand.

Bowling, M., Browning, B., Chang, A., and Veloso, M. (2004). Plays as team plans for coordination and adaptation. In Polani, D., Browning, B., Bonarini, A., and Yoshida, K., editors, *RoboCup 2003: Robot Soccer World Cup VII*, volume 3020 of *Lecture Notes in Computer Science*, pages 686–693. Springer Verlag, Berlin, Germany.

Castelpietra, C., Iocchi, L., Nardi, D., Piaggio, M., Scalzo, A., and Sgorbissa, A. (2001). Communication and coordination among heterogeneous mid-size players: ART99. *Lecture Notes in Computer Science*, 2019:86–95.

Khatib, O. (1985). Real-time obstacle avoidance for manipulators and mobile robots. In *Proceedings of the IEEE International Conference on Robotics and Automation*, pages 500–505.

Kitano, H., Asada, M., Kuniyoshi, Y., Noda, I., and Osawa, E. (1997). RoboCup: The robot world cup initiative. In Johnson, W. L. and Hayes-Roth, B., editors, *Proceedings of the First International Conference on Autonomous Agents (Agents'97)*, pages 340–347, New York. ACM Press.

Koren, Y. and Borenstein, J. (1991). Potential field methods and their inherent limitations for mobile robot navigation. In *Proceedings of the IEEE International Conference on Robotics and Automation*, pages 1398–1404, Sacramento, CA.

Laue, T. and Röfer, T. (2004). A behavior architecture for autonomous mobile robots based on potential fields. In *8th International. Workshop on RoboCup 2004 (Robot World Cup Soccer Games and Conferences), Lecture Notes in Artificial Intelligence*, Lecture Notes in Computer Science, Berlin, Germany. Springer Verlag.

Roth, M., Vail, D., and Veloso, M. (2003). A real-time world model for multi-robot teams with high-latency communication. In *Proceedings of the IEEE/RSJ International Conference on Intelligent Robots and Systems*, volume 3, pages 2494–2499.

Uther, W., Lenser, S., Bruce, J., Hock, M., and Veloso, M. (2002). CM-Pack'01: Fast legged robot walking, robust localization, and team behaviors. In Birk, A., Coradeschi, S., and Tadokoro, S., editors, *RoboCup 2001: Robot Soccer World Cup V*, Lecture Notes in Computer Science, pages 693–696. Springer Verlag, Berlin, Germany.

Vail, D. and Veloso, M. (2003). Dynamic multi-robot coordination. In *Multi-Robot Systems: From Swarms to Intelligent Automata, Volume II*, pages 87–100. Kluwer Academic Publishers.

Veloso, M., Stone, P., and Bowling, M. (1999). Anticipation as a key for collaboration in a team of agents: A case study in robotic soccer. In *Proceedings of SPIE Sensor Fusion and Decentralized Control in Robotic Systems II*, volume 3839, Boston.

Weigel, T., Auerbach, W., Dietl, M., Dümler, B., Gutmann, J.-S., Marko, K., Müller, K., Nebel, B., Szerbakowski, B., and Thiel, M. (2001). CS Freiburg: Doing the right thing in a group. *Lecture Notes in Computer Science*, 2019:52–63.

PARALLEL STOCHASTIC HILL-CLIMBING WITH SMALL TEAMS

Brian P. Gerkey, Sebastian Thrun
Artificial Intelligence Lab
Stanford University
Stanford, CA 94305, USA
gerkey@ai.stanford.edu, thrun@stanford.edu

Geoff Gordon
Center for Automated Learning and Discovery
Carnegie Mellon University
Pittsburgh, PA 15213, USA
ggordon+@cs.cmu.edu

Abstract We address the basic problem of coordinating the actions of multiple robots that are working toward a common goal. This kind of problem is NP-hard, because in order to coordinate a system of n robots, it is in principle necessary to generate and evaluate a number of actions or plans that is exponential in n (assuming $P \neq NP$). However, we suggest that many instances of coordination problems, despite the NP-hardness of the overall class of problems, do not in practice require exponential computation in order to arrive at good solutions. In such problems, it is not necessary to consider all possible actions of the n robots; instead an algorithm may restrict its attention to interactions within small teams, and still produce high-quality solutions.

We use this insight in the development of a novel coordination algorithm that we call *parallel stochastic hill-climbing with small teams*, or *Parish*. This algorithm is designed specifically for use in multi-robot systems: it can run off-line or on-line, is easily distributed across multiple machines, and is efficient with regard to communication. We state and analyze the Parish algorithm present results from the implementation and application of the algorithm for a concrete problem: multi-robot pursuit-evasion. In this demanding domain, a team of robots must coordinate their actions so as to guarantee location of a skilled evader.

Keywords: coordination, multi-robot systems, pursuit-evasion

L.E. Parker et al. (eds.),
Multi-Robot Systems. From Swarms to Intelligent Automata. Volume III, 65–77.

1. Introduction

Multi-robot systems have the potential to be far more useful than single robots: multiple robots may perform a given task more efficiently than a single robot, multiple robots may be more robust to failure than a single robot, and multiple robots may be able to achieve tasks that are impossible for a single robot. However, reaching that potential can be extremely difficult, especially in the case where multiple robots make task achievement *possible*, rather than simply *better*. The difficulty arises primarily from the combinatorial possibilities inherent in the problem of coordinating the actions of multiple robots, which is in general NP-hard ((Garey and Johnson, 1979)). Given a system of n robots and a common goal, it may be necessary to generate and evaluate a number of actions or plans that is exponential in n (assuming that $P \neq NP$).

One common way to attack such a problem is brute-force search in the joint state/action space. That is, treat the multi-robot system as one many-bodied robot and look through the exponentially many possibilities until the right answer is found. Though this approach will produce an optimal solution, it is only viable on simple problems, as the necessary computation quickly becomes intractable as the number of robots and/or the complexity of the problem grows. This fact contradicts the intuition that having more robots available should make a task easier, rather than harder, to solve. Additionally, this approach is undesirable for most robotic applications, because it requires a centralized planner / executive, which precludes local control decisions at the level of an individual robot.

Another, more popular, approach is to treat the multi-robot system as a collection of independent single robots and allow each one to make individual control decisions, irrespective of the other robots' actions. This approach scales very well, as it requires each robot to consider only its own possible actions, the number of which remains constant as the number of robots grows. Unfortunately, this technique will not necessarily produce a good solution. In fact, if the actions of the robots *must* be coordinated in order to achieve a task, then allowing them to simply make individual choices without considering or consulting each other is unlikely to lead to any solution at all.

We believe that between these two extremes lies fertile ground for the development of heuristic multi-robot coordination algorithms that produce good solutions yet scale well with the number of robots. In particular, we suggest that many multi-robot problems can be solved quickly and effectively by allowing the formation of and planning for small teams over short time horizons. That is, rather than considering the possible actions of all n robots or of just 1 robot, consider groups of up to t robots, where $1 \leq t \leq n$, but prefer smaller groups, because they are computationally cheaper to coordinate. In this paper we introduce an algorithm, *parallel stochastic hill-climbing with small teams*, or

Parish, which combines the idea of small teams with the use of heuristics and stochastic action selection. In addition to scaling well and tending to produce good solutions to coordination problems, Parish is easily distributable, and can be executed either on-line or off-line, both of which are desirable properties for multi-robot algorithms.

We have implemented Parish for the problem of *multi-robot pursuit-evasion*, in which a group of robots must work together to search a given environment so as to guarantee location of a skilled mobile evader. This is a difficult problem that clearly requires coordination among the robots (a single robot is only capable of clearing environments that are topologically equivalent to a single hallway). And, unlike more weakly interactive tasks, like foraging, pursuit-evasion occasionally requires very tight coordination between robots in order to make any progress at all. We provide results from tests in simulation of search strategies produced by Parish.

2. Background and related work

The first rigorous formulation of the pursuit-evasion problem is due to Parsons, who restricted his study to the case in which the environment is a discrete graph ((Parsons, 1976)). Nothing is known about the location or motion of the evader, who is assumed to be able to move arbitrarily fast through the graph. The evader can occupy any edge in the graph; to find the evader, a searcher must walk along the edge occupied by the evader and "touch" the evader. The entire graph is initially *contaminated*, which means that the evader could be anywhere. As the search progresses, an edge is *cleared* when it is no longer possible for the evader to occupy that edge. Should it later happen that the evader could have moved back to a previously clear edge, that edge is said to be *recontaminated*. Using this terminology, the goal of the problem can be restated as follows: find a trajectory for each searcher such that the an initially contaminated graph is cleared.

More recently, a visibility-based version of the pursuit-evasion problem was introduced ((Suzuki and Yamashita, 1992)), which changed the domain from discrete graphs to continuous polygonal free spaces. Complete algorithms have been described for searchers having either 1 or 2 "flashlights" ((Lee et al., 2002)), omnidirectional vision ((Guibas et al., 1999)), and limited field-of-view vision ((Gerkey et al., 2004)). Randomized pursuit algorithms have also been studied, in both discrete graphs ((Adler et al., 2003)) and polygonal free spaces ((Isler et al., 2003)).

3. Algorithm

The *Parish* algorithm coordinates a multi-robot system in a scalable manner by considering the possible actions of not only single robots, but also small teams of robots. The general form of the algorithm can be summarized as follows:

Algorithm *Parish: Parallel stochastic hill-climbing with small teams*
Input: n robots; multi-robot problem M; maximum team size $t \leq n$; value heuristic $v(q)$; probability distribution $P(q)$, $P(q_j) > P(q_i) \Leftrightarrow v(q_j) \geq v(q_i)$

```
while M not done
    do parallel for each robot s
        do for l ← 1 to t
            do Q_l ← {q : q is a feasible l-searcher plan involving s} ∪ {∅}
               Sample q̂_l from Q_l according to P(q)
               if q̂_l ≠ ∅
                 then Execute q̂_l
                      break
```

The *value* heuristic $v(q)$ has two components: a *benefit* heuristic $b(q)$ and a *cost* function $c(q)$. The benefit heuristic b estimates the (possibly negative) marginal benefit (i.e., progress that would be made toward solution) of a given plan. In other words, b estimates the optimal value function, which is unknown (computing the optimal value function is equivalent to solving the original *NP*-hard problem). If a plan q involves any robots that are currently part of other teams that are engaged in other plans, then $b(q)$ includes an estimate of the (probably negative) benefit that will result from disbanding those teams and halting the execution of those other plans. The function c calculates, in the same units as b, the cost of executing a given plan. This cost can be any salient aspect of the domain that is external to progress, such as distance moved. The *value* of a plan q is then $v(q) = b(q) - c(q)$.

Because the heuristic b is only an *estimate* of the true benefit of a given plan, we cannot always select the highest-valued plan. Such a strategy will, in all but the simplest problems, lead to local maxima of progress from which the system will not escape. Thus we employ a stochastic selection rule: rather than greedily selecting the apparently best plan, we sample a plan $\hat{q}_l$ from the set Q_l of available plans, according to a probability distribution $P(q)$ that prefers higher-valued plans but sometimes selects an apparently worse plan. This technique is commonly used in optimization to escape from local extrema and is in reinforcement learning to balance exploration against exploitation. So robots executing Parish are collectively hill-climbing according to local progress gradients, but stochastically make lateral or downward moves to help the system escape from local maxima.

The exact nature of the selection rule can be adjusted according to the accuracy of the benefit heuristic. If b is known to be a very accurate estimate of the optimal value function, then the highest-valued plan should be selected with accordingly high probability, and vice versa if b is known to be less accurate (of course, if b is very inaccurate, then progress will be slow, and more effort should likely be put toward designing a better heuristic).

Since the robots make plans individually, the computation of the algorithm can easily be distributed across multiple machines, with communication required only to update the state of the problem and to form (or break up) teams. If a good model of the environment is available, then Parish can run off-line, with the robots interacting with this model to produce a plan for later execution. If no good model is available, or if the environment is dynamic, then Parish can run on-line, with the robots interacting with the environment directly. Also, robots will tend select and execute single-robot plans, if good ones can be found, because they do not require breaking up other teams. Thus they will make individual progress as long as possible, until such time as team formation is more beneficial.

3.1 Economic interpretation

As is the case with many multi-agent search algorithms, there is an obvious economic interpretation of Parish. The multi-robot system can be seen as a synthetic economy, in which individual robots can buy the services of other robots. A robot receives (possibly negative) "reward" for making (possibly backward) progress toward the goal. Each robot then selfishly tries to "earn" as much reward as possible. The value, $v = b - c$, that a robot attaches to a plan that it has formulated is the "price" that that robot will "pay" in order to form the team that will help in executing the plan (the robot may offer a price slightly less than v, in order to retain some positive profit). A robot only joins a team when it is offered a sufficiently high price to take it away from its current team, if any. Stochastic plan selection then corresponds to a robot occasionally making a choice that does not maximize its reward, to account for the fact that, because of inaccuracies in prices (i.e., values), strict reward-maximization will not necessarily lead to a solution.

Although this economic interpretation relates our algorithm to previous work in economically-inspired multi-robot coordination approaches (e.g., (Gerkey and Matarić, 2002, Dias and Stentz, 2003)), we do not find it particularly helpful. Coordination algorithms such as Parish can be understood and clearly stated as instances of distributed search or optimization; economic interpretations can unnecessarily cloud the discussion by introducing misleading analogies between synthetic markets as used by robots and real markets as used by humans.

3.2 Application to multi-robot pursuit-evasion

We now make Parish concrete by explaining how we apply it to the problem of multi-robot pursuit-evasion and stating the resulting algorithm. In the multi-robot pursuit-evasion problem, a team of n robots is required to search an environment (of which a map is provided) so as to guarantee location of a

Figure 1. *The Botrics Obot mobile robot, equipped with a SICK scanning laser range-finder, which has a 180° sensor field.*

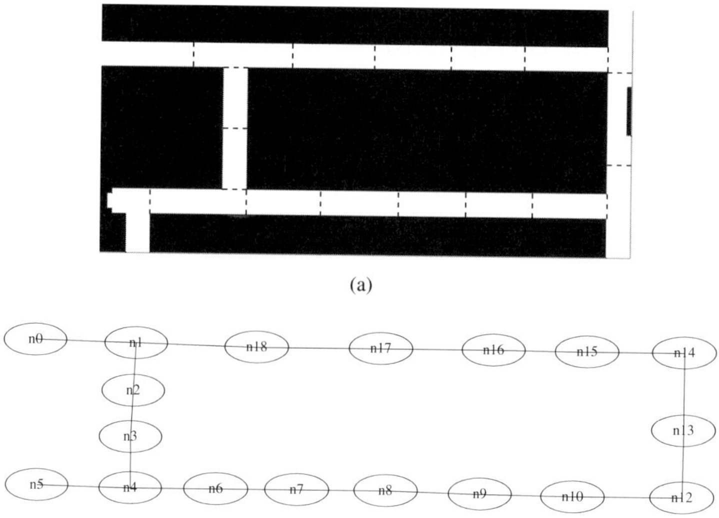

(b)

Figure 2. *An office-like environment, decomposed into convex regions (a) and then transformed into a discrete graph (b).*

skilled mobile evader. The only information available about the evader is its size and maximum speed; no model of its initial position or subsequent trajectory is given. For our purposes, a robot "finds" the evader if the evader is detected within the robot's sensor field. Our robots are each equipped with a scanning laser range-finder that provides a 180° field of view and reliable detection range of approximately 8 meters (Figure 1).

We first transform our problem to an instance of Parsons's discrete graph search problem ((Parsons, 1976)). This transformation involves decomposing the free space in the given map into finitely many regions such that a single robot can clear a region by standing anywhere on and perpendicular to the region border, while looking into the region. Furthermore, we want to guarantee that a differential-drive robot with a 180° field of view can move from one border of a region to any other border of the same region and keep the destination border in view along the way. Two necessary and sufficient conditions for the regions are that they each: (i) be convex, and (ii) have no dimension greater than the maximum sensor range (8 meters). For the work presented in this paper, the decomposition was performed manually, but the process could be automated according to visibility constraints (e.g., (Guibas et al., 1999, Gerkey et al., 2004)). Given such a region decomposition, we construct an undirected graph $G = (V, E)$, where the vertices V are the regions, and the edges E are the borders where adjacent regions meet. An example decomposition and the resulting graph are shown in Figure 2.

We can then apply Parish, stated below, to the graph G, and transform the resulting solution back to the robots' control space, with each move in the graph becoming a move to a region border in the physical environment.

Preliminaries:

- Searcher positions and edge contamination states are stored as labels in the graph.
- The graph, the list of teams, and the list of plans are shared data structures: each searcher has an identical copy of each structure, and a mutual exclusion mechanism is used to ensure consistency when making changes.
- S_i denotes searcher i.
- Given a list L, $L[i]$ denotes the ith element of L.
- A *plan* q specifies a sequence of moves for one or more searchers.
- The *null plan*, denoted $\emptyset$, makes no moves.
- Given a plan q, q.members() returns the set of searchers required to execute q.
- $G' \leftarrow G + q$ denotes the application of plan q to graph G to produce the resulting graph G'.
- Given a team T with n members, to *disband* T is to separate the members of T into n singleton teams, one individual per team.

Algorithm *Parish for multi-robot pursuit-evasion*
Input: Connected, undirected graph G; n searchers placed in G (if initial placement is not given, place them randomly); maximum team size t; value heuristic $v(G,q)$; probability distribution $P(q)$, $P(q_j) > P(q_i) \Leftrightarrow v(q_j) \geq v(q_i)$

$T \leftarrow [\,]$ (∗ List of teams ∗)
$A \leftarrow [\,]$ (∗ List of plans ∗)
for $i \leftarrow 1$ **to** n
    do (∗ Start with singleton teams and no plans ∗)
        T.append($\{S_i\}$)
        A.append($\emptyset$)
while not done
    do (∗ Each team decides what to do, in parallel ∗)
        parallel for $i \leftarrow 1$ **to** len(T)
            do if $A[i] = \emptyset$
                then (∗ No plan, so this team has only one member; call it s ∗)

```
            s ← s : s ∈ T[j]
            (∗ Consider teams of increasing size, up to t ∗)
            for l ← 1 to t
                do (∗ Make some l-searcher plans, but also consider the null plan ∗)
                    Q_l ← {q : q is a feasible l-searcher plan involving s} ⋃ {∅}
                    Sample q̂_l from Q_l according to P(q)
                    if q̂_l = ∅
                      then (∗ We chose the null plan; keep looking ∗)
                            continue
                      else (∗ Assemble the team, maybe disbanding other teams ∗)
                            for j ← 1 to len(T), j ≠ i
                                do for r ∈ q̂_l.members()
                                        do if r ∈ T[j]
                                              then Disband T[j]
                                                   T[i] = T[i] ⋃ r
                            (∗ Store the chosen plan and begin executing it ∗)
                            A[i] ← q̂_l
                            G ← G + first step of A[i]
                            (∗ We have a satisfactory plan; stop looking ∗)
                            break
        else (∗ We already have a plan, so keep executing it ∗)
            G ← G + next step of A[i]
            if just executed last step of A[i]
              then (∗ This team has finished its plan; disband it ∗)
                    Disband T[i]
```

4. Results

We implemented Parish as stated in the previous section and tested it on several environments. The tests were carried out using Stage, a sensor-based multi-robot simulator; experience has shown that results in Stage can be reliably replicated with with physical (indoor, planar) robots ((Gerkey et al., 2003)). Animations can be found at: **http://ai.stanford.edu/~gerkey/research/pe/**.

The benefit heuristic b is the (possibly negative) number of regions that would be cleared by executing a given plan. The cost function c is proportional to distance traveled during a given plan, calculated as number of regions traversed. The maximum team size is $t = 2$, and the robots are restricted to making plans that move each team member once. Specifically, each robot S_i only considers plans of the following form:

- (team size 1) Move S_i to an adjacent region.
- (team size 2) Move another robot S_j ($i \neq j$) to cover the region currently covered by S_i, then move S_i into an adjacent region.

The stochastic selection rule is ε-greedy, in which the highest-valued plan is selected with probability $(1 - \varepsilon)$, and otherwise one of the remaining options is chosen with uniform probability. For the results presented here, $\varepsilon = 0.1$. We

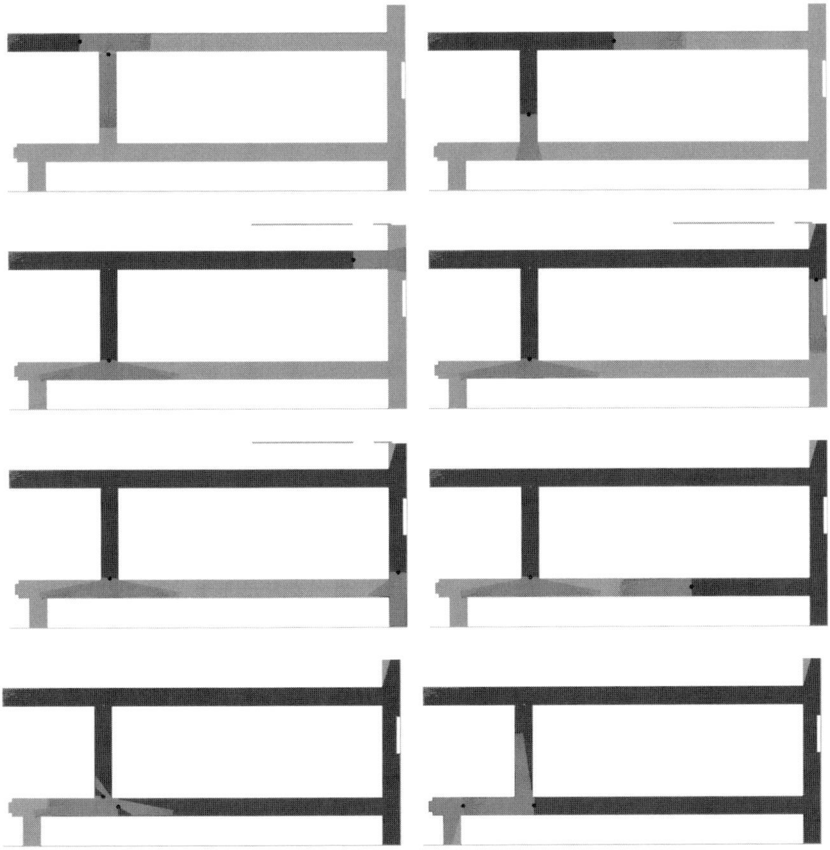

Figure 3. *(In color where available). Two robots searching an office-like environment. Black circles represent robots; blue areas are clear; red areas are in view; and purple areas are contaminated (i.e., the evader may be hiding there).*

assume the environment is static, and so are free to run Parish off-line, then execute the resulting plan with the simulated robots.

Interestingly, adding just this limited and myopic coordination is sufficient to produce good solutions. For example, shown in Figure 3 are snapshots from a run with 2 robots in an office-like environment. As can be seen in that figure, the robots cooperate to clear the environment quite efficiently, without allowing recontamination. In fact, Parish reliably produces solutions for this and similar environments that are optimal in the total path length. (we compute optimal solutions using brute-force A* search in the joint action/state space of all the robots).

The effect of small-team coordination can be clearly seen in Figure 4, taken from a simulation run in which 5 robots work together to clear one floor of an office building, using a sensor-based map. In this sequence, a 2-robot plan calls for the robot initially at the lower right to move up and block the central open area so that another robot can move left and keep searching. Without such interactions, the robots are not capable of clearing this complex environment.

5. Summary and future work

We introduced the *Parish* algorithm, which allows for scalable and efficient coordination in multi-robot systems. The key insight of the algorithm is that the combination of small teams, simple heuristics, and stochastic action selection can be extremely effective in solving otherwise difficult multi-robot problems. Our algorithm is easily distributable and can run on-line or off-line, making it especially suitable for use in physical robots systems. We presented results from simulation that demonstrate the efficacy of Parish in coordinating robots engaged in a pursuit-evasion task.

Our current work on this algorithm follows 3 paths. First, we are moving to physical robots, where Parish will run on-line, and fully distributed. Second, we are rigorously analyzing Parish and comparing it to competitor algorithms, such as non-cooperative greedy, and centralized A*. It will be important to establish the average-case and worst-case performance of Parish, in terms of solution quality and computational requirements (i.e., amount of the search space that is actually explored), as compared to existing alternatives (Figure 5). Finally, we are applying Parish to other multi-robot coordination problems.

References

Adler, M., Räcke, H., Sivadasan, N., Sohler, C., and Vöcking, B. (2003). Randomized Pursuit-Evasion in Graphs. *Combinatorics, Probability and Computing*, 12(3):225–244.

Dias, M. B. and Stentz, A. (2003). TraderBots: A Market-Based Approach for Resource, Role, and Task Allocation in Multirobot Coordination. Technical Report CMU-RI-TR-03-19, Robotics Institute, Carnegie Mellon University, Pittsburgh, Pennsylvania.

Garey, M. R. and Johnson, D. S. (1979). *Computers and Intractability: A Guide to the Theory of NP-Completeness*. W. H. Freeman.

Gerkey, B. P. and Matarić, M. J. (2002). Sold!: Auction methods for multi-robot coordination. *IEEE Transactions on Robotics and Automation*, 18(5):758–768.

Gerkey, B. P., Thrun, S., and Gordon, G. (2004). Visibility-based pursuit-evasion with limited field of view. In *Proc. of the Natl. Conf. on Artificial Intelligence (AAAI)*, pages 20–27, San Jose, California.

Gerkey, B. P., Vaughan, R. T., and Howard, A. (2003). The Player/Stage Project: Tools for Multi-Robot and Distributed Sensor Systems. In *Proc. of the Intl. Conf. on Advanced Robotics (ICAR)*, pages 317–323, Coimbra, Portugal.

Guibas, L. J., Latombe, J.-C., LaValle, S. M., Lin, D., and Motwani, R. (1999). A Visibility-Based Pursuit-Evasion Problem. *Intl. J. of Computational Geometry & Applications*, 9(4 & 5):471–493.

Isler, V., Kannan, S., and Khanna, S. (2003). Locating and capturing an evader in a polygonal environment. Technical Report MS-CIS-03-33, Dept. of Computer Science, Univ. of Pennsylvania.

Lee, J.-H., Park, S.-M., and Chwa, K.-Y. (2002). Simple algorithms for searching a polygon with flashlights. *Information Processing Letters*, 81:265–270.

Parsons, T. (1976). Pursuit-evasion in a graph. In Alavi, Y. and Lick, D., editors, *Theory and Applications of Graphs*, Lecture Notes in Mathematics 642, pages 426–441. Springer-Verlag, Berlin.

Suzuki, I. and Yamashita, M. (1992). Searching for a mobile intruder in a polygonal region. *SIAM J. on Computing*, 21(5):863–888.

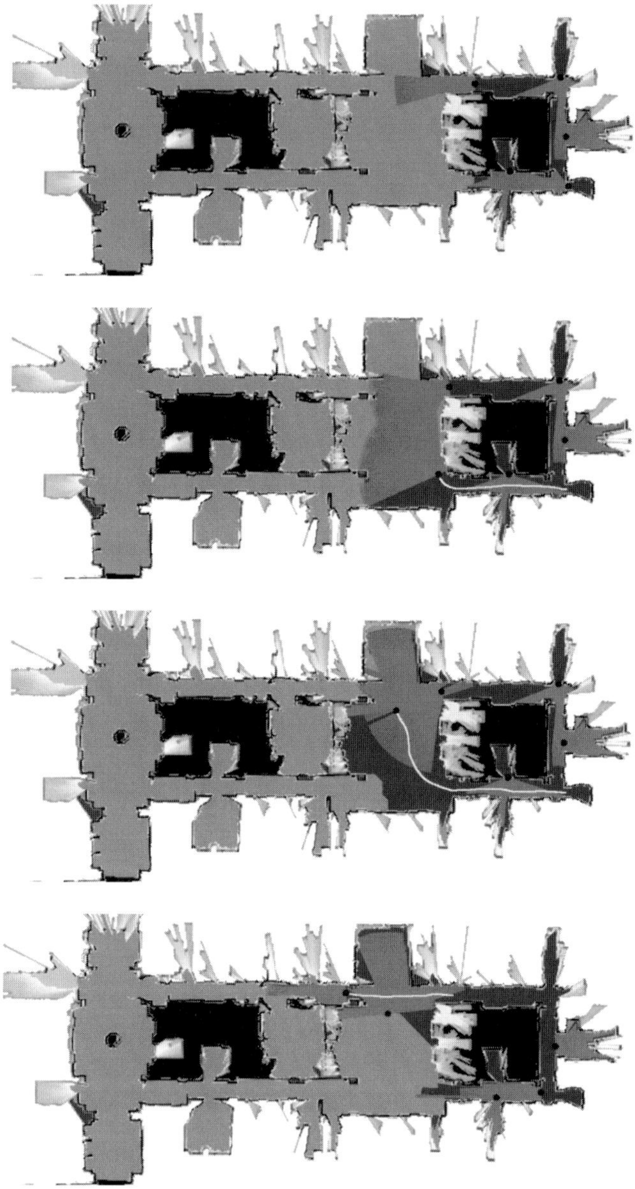

Figure 4. (In color where available). An example of small-team coordination taken from a test in which 5 robots cleared a large building. As part of a 2-robot plan, the robot that is initially in the lower right corner moves up and left to block the central open area so the robot that another robot can move left and keep searching.

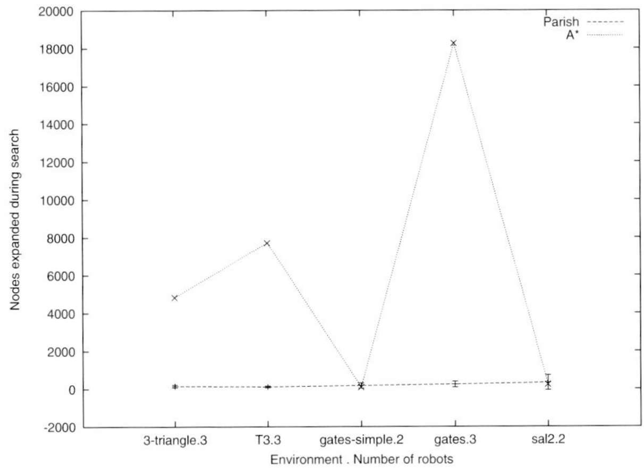

(a)

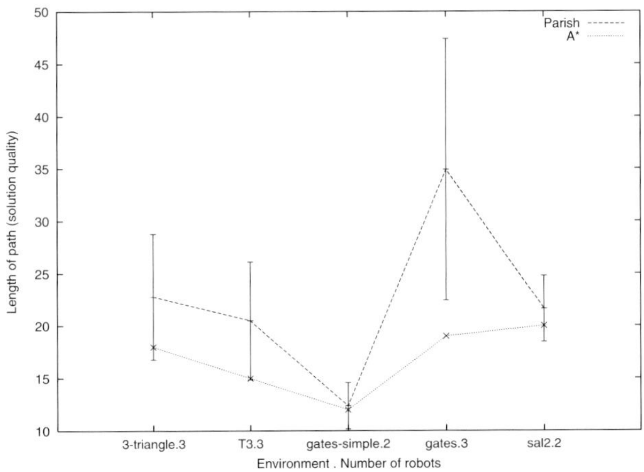

(b)

Figure 5. *Comparison of Parish and A* in planning pursuit strategies in various environments. Shown in (a) is the number of nodes expanded during the search, and in (b) is the length of the solution found (smaller is better in both cases). Results for Parish, which is stochastic, show the experimental mean and standard deviation, computed from 100 runs in each environment.*

TOWARD VERSATILITY OF MULTI-ROBOT SYSTEMS

Colin Cherry and Hong Zhang
Department of Computing Science
University of Alberta
Edmonton, Alberta Canada T6G 2E8
{colinc, zhang} @cs.ualberta.ca

Abstract This paper provides anecdotal evidence that the group behavior exhibited by a collective of robots under reactive control is as much due to the design of their internal behaviors as to the external conditions imposed by the environment in which the robot collective operate. Our argument is advanced through the examination of a set of well-known collective robotics tasks that involve, in one form or another, the movement of materials, namely, foraging, transport, and construction. We demonstrate that these seemingly different tasks can all be achieved by one controller that employs behaviors identical across the task domains but parameterized with a couple of task-specific constants. The implementation of the study is conducted with the help of the TeamBots robot simulator.

1. Introduction

Collective robotics is concerned with the use of multiple robots for the performance of tasks that are inefficient, difficult or impossible by robot singletons (Balch and Parker, 2002). A dominant and successful design approach to multi-robot systems (MRS) is based on behavior-based control (Brooks, 1986), which uses a hierarchy of simple reflexive or reactive behaviors at the level of individual robots whose interaction with each other as well as with the environment can lead to the emergence of desired group-level behaviors. Behavior-based control can be implemented with the well-known subsumption architecture (Brooks, 1986) or motor schemas (Arkin, 1998), which is adopted in this study. Numerous successful studies have clearly established the validity of this design approach.

One of the cornerstones of behavior-based robotics is the recognition of the importance of the environment on the behaviors displayed by a robot that interacts with it. Given the complexity and variability of the world in which a robot must operate, rather than attempting to model the world exactly and formulate

L.E. Parker et al. (eds.),
Multi-Robot Systems. From Swarms to Intelligent Automata. Volume III, 79–90.

plans and actions, a behavior-based approach uses carefully designed rules that sense and react to the changes in the world, and this approach has proved to be more robust and effective in many situations than the traditional deliberative control (Gat, 1997).

In this paper, we argue that the importance of the environment can be taken one step further, in the sense that the external behaviors displayed by a single robot or a multi-robot collective can, to a significant extent, be attributed to the environment in which the robots reside. We will support this argument by demonstrating that only minimum changes need to be made to the robot controller when the robot system is confronted with a seemingly different task. Specifically, we will use a set of three well-known collective robotics tasks which involve, in one form or another, the movement of materials, namely, foraging, group transport (box-pushing), and collective construction. We will show that, to solve all three tasks, structural change to the robot controller is unnecessary, and that only parameters that drive the internal behaviors of the robots need to be adjusted. This observation represents a further development beyond the assertion that basis behaviors exist that are capable of wide variety of tasks (Mataric, 1995).

1.1 Collective Robotic Building Tasks

The work in (Balch and Arkin, 1995) describes in detail the foraging task and the corresponding reactive foraging controller. A number of "attractors" litter the field, and must be brought to a home position. The attractors have mass, and heavy attractors can be carried more quickly with more than one robot working together. The robots can communicate, and therefore have the ability to signal other robots to work on the same attractor as them.

Group transport, or box pushing, describes any task where robots must work together to move one object that a single robot can not move alone. The task modeled in this paper is most similar to the tasks investigated in (Kube and Zhang, 1993) and (Kube and Zhang, 1997), which tackle the problem of having robots push an attractor to a specific destination. Their work uses robots controlled by hierarchical FSAs to push a circular box to a lit home area. The robots are able to see both the home area and the attractor, allowing them to determine if the attractor is between them and the goal. This notion of positioning so that the attractor is between the robot and the destination will become central to the pushing strategy used later in this paper.

The problem of collective construction is studied in (Parker et al., 2003). Robots are initially positioned in a clear area surrounded by rubble, and they proceed to clear the debris to create a larger, circular nest, where the nest walls are composed of the rubble. Inspired by construction by ants, (Parker et al., 2003) takes an extreme minimalist approach to collective construction, called blind bulldozing. The robots being controlled in this paper, however, have access to considerably more sensing power than those used in (Parker et al., 2003), with the ability to sense attractors, sense a home position, and to differentiate between those attractors that are in place and those that are not.

1.2 Problem Statement

The goal of our study is to design and test a single controller capable of executing all three tasks described above. All three tasks involve moving some number of objects from their current location to a destination. The major variables in these tasks are whether or not the robots are capable of moving individual objects alone, and the destination for the objects. Regardless of whether robots are working alone or in teams, pushing is always used as the method of moving objects.

The robots' physical capabilities and sensors remain static through all three tasks. The only changes allowed from task to task are those in the external environment, and adjustments to some small number of controller parameters. These parameters are intended to make small, task-specific changes to the controller: adjustments to schema weights or polarities, or to the triggers for perceptual cues. Once designed, the controller will be implemented in the TeamBots simulator, and tested on the three tasks.

2. Design of Versatile Controller

This section will describe the multitask controller used to conduct foraging, group transport and nest construction tasks. Our control system will follow the motor schema approach popularized by (Arkin, 1998). This approach uses a perception-triggered finite state machine (FSM) to switch the robot between states, corresponding to the major steps needed for any task.

We will begin by describing the general strategy employed in any pushing task and the corresponding finite state machine. The behavior in each state will then be described in terms of combinations of base behaviors. We will then describe the perceptual cues that trigger state transitions. We end this section with a discussion of the three parameters that are used to adjust the generic system to favor particular tasks: the cooperation and positioning parameters (c and p), and the building flag (b).

2.1 Generic Strategy and Finite State Machine

Viewed at a high level, the generic pushing robots tackle all material transport problems with the same strategy: they must find an object to push, get into a good pushing position, and then push object to its destination. Starting with a foraging FSM that is similar to the one used in (Balch and Arkin, 1995), we make small modifications to reflect the differences between the generic

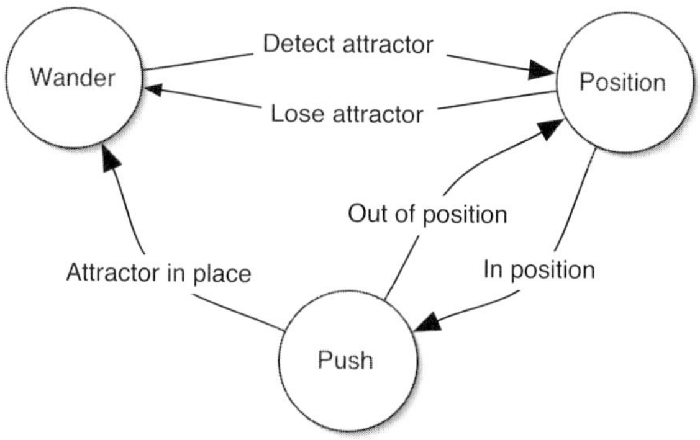

Figure 1. The finite state machine for the generic pushing robot.

pushing task and foraging if a gripper is available. The resulting controller is pictured in Figure 1.

The purpose of the "Wander" state is to find an attractor to work on. Wander produces a random walk to perform this attractor search, consisting of the four schemas. *Noise* produces random motion. *Avoid obstacles* keeps robots from colliding, and ensures they do not redundantly search the same area. *Swirl obstacles to noise* ensures that the robot takes a smooth route around obstacles in its current direction of travel. Finally, *Draw to home*, active only when the build flag, to be described later, is set, encourages the robot to explore an area close to home. This is the only instance where an extra schema was added to a state for a particular task. As with any schema-based control, the outputs of the behaviors (motor schemas) are combined in the form of a weighted sum to produce the final actuator commands (Arkin, 1998).

The purpose of the "Position" state, not to be confused with the position flag p, is to bring the robot into a good position for pushing the attractor to its destination. Position causes the robot to move in a circular path around the robot's closest, targeted attractor, without necessarily coming into contact with the attractor. It consists of the five schemas, including the *Swirl obstacles to noise* and *Noise* schemas as in the Wander state. The first of the three new schemas, *Swirl target to home*, produces a circular motion so the robot moves around its closest attractor. *Move to target* schema moves the robot toward the target, so it is less likely to lose sight of the target while rotating around it. *Swirl obstacles to home* schema ensures a smooth path around other robots in the current direction of travel. Finally, *Watch target* is active for the robot's turret, always drawing it toward the current target, so the robot does not lose sight of it while rotating.

The purpose of "Push" is to have the robot to move to a destination point, pushing an attractor in front of it. "Push" consists of the same set of schemas as "Position", with the exception of *Swirl obstacles to target*, in place of *Swirl obstacles to home*, which ensures smooth motion around an obstacle in the robot's current direction of travel.

2.2 Perceptual Cues

Transitions between states depend on a set of perceptual cues described in this section. First of all, the "detect attractor" cue is triggered when an attractor object falls within the robot's visual sensor range. Similarly, "lose attractor" is triggered if no attractors are currently in visual range.

Secondly, as in previous studies (Balch and Arkin, 1995), we assume that the robot is constantly able to sense its home. For foraging and group transport, the "in position" cue is triggered when the robot's closest visible attractor lies between the robot and home. Whether or not an attractor is between the robot and its home is determined by the alignment between vector from the robot to the attractor and that from the robot to *Home*.

The betweenness threshold for "out of position" is set to be more tolerant than that of "in position". This produces behavior where the robot is very picky about when it starts to consider itself in position, but then allows for a fair amount of leeway before it actively corrects its position. This prevents rapid and ineffectual switching between the "Position" and "Push" states. When the build parameter is set to produce building behavior, this perceptual cue is inverted. In this case, it detects when the robot lies between home and the attractor. See Section 2 for more details.

2.3 Task Parameters

The goal of this research is to construct a controller that is equally switches gracefully among the tasks of foraging, group transport, and nest construction, under the influence of only external conditions defined by the environment. This will be accomplished with the help of the three simple parameters described below, which are devised in order to tune the robots to be predisposed to perform a specific task. One should note, though, that structure of the controller does not change and that they affect primarily schema weights and the direction of certain forces. In other words, the parameters have been designed to affect the performance of the controller while maintaining its spirit.

Cooperation: c. The cooperation parameter c determines how well robots work together, and conversely, how carefully they avoid redundant effort. It varies in value from 0 to 1, with $c = 1$ indicating maximum cooperation. It affects only the *Avoid obstacle* and *Swirl obstacle* schemas in the "Position" and "Push" states.

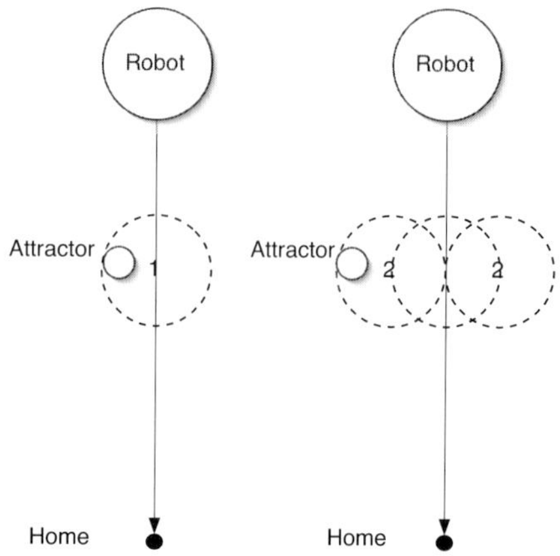

Figure 2. An example of pushing position with respect to robot radii

In "Position," the avoid schemas have their weight set according to $2(1-c)$. The result is that when cooperation is low, a robot will be unlikely to reach a good pushing position (and thus enter "Push") for an attractor that already has another robot near by. When cooperation is high, several robots will be able to target the same attractor easily.

Conversely, in "Push", the avoid schemas have their weight set directly according to c. As is described above, robots need to avoid each other when cooperating to push a large object, or no room will be made for potentially helpful robots. When cooperation is low, robots will ignore each other, and concentrate on pushing. This rarely results in two robots "fighting over" the same object, though, as they are unlikely to both target the same object due to the high avoid weight while in the "Position" state.

Position: p. The position parameter p intuitively determines how fussy a robot is about positioning itself before it starts pushing. A lower p indicates that the robot requires a position that is closer to the optimal position before it starts pushing. Numerically, p, is the number of robot radii away an attractor is allowed to be from the point where it would be exactly on the path between the robot and its destination. For example, take the two situations pictured in Figure 2. In the situation on the left, a robot with $p=1$ would transition into "Push," as would a robot with $p=2$. However, in the situation on the right, only a robot with $p=2$ would transition into "Push." Note that the transition happens regardless of the object radius, as a robot may not be able to tell an object's true radius in practice.

Build Flag: b. The build flag switches the robot from a foraging goal to a nest building goal, pushing objects out and away from a central point, instead of pushing them in toward the same point. It has the most dramatic effect on the controller of all the parameters, but it is our hope that though its effect may be large, it is conceptually a very small change: it simply reverses the polarity of the robots' home.

When the build flag is set, all schemas that would move the robot toward home become moves away from home and vice-versa. This affects only the *Swirl X to home* schemas in the "Position" and "Push" states. When foraging, these schemas swirl the robot away from home, so that the robot keeps the attractor between itself and home, and therefore pushes the attractor to home. When constructing, these schemas swirl the robot toward home, so it stays between home and the attractor. Therefore, it pushes the attractor away from home. To agree with the above changes, the build flag also redefines the "(Still) In position" perceptual cue, so that the robot checks if it is between home and the attractor when building. Similarly, it redefines the robots' attractor filter so that all attractors that are a certain distance away from home are filtered out when building, as opposed to filtering attractors sufficiently close to home when foraging.

Finally, in the only change that is not directly tied to reversing the polarity of home, it adds an extra schema to the "Wander" state, causing the robot to stay close to home, and to avoid wandering in the portions of the map where nothing needs to be done. This also has the visually pleasing effect of having the robots peacefully "live" inside their nest after construction is complete.

Initial parameterizations for specific tasks. The parameters were created with the intention of easily characterizing the three target tasks with parameter values. Foraging small, light attractors was envisioned as a low-cooperation foraging task, and therefore was characterized with the parameter settings $c = 0$, $p = 1$ and $b = false$. Group transport was seen as a high-cooperation foraging task, and was characterized with $c = 1$, $p = w$ and $b = false$, where w is the minimum number of robots needed to move the heaviest attractor. Finally nest construction with small, light attractors was seen as a low-cooperation building task, and was characterized with $c = 0$, $p = 1$ and $b = true$.

3. Experimental Design

We have conducted experiments for the purpose of both proof of concept and tests for parameter sensitivity. Only the results in the first category are described due to the limit on the length of the paper. Proofs of concept are designed to answer the question of whether or not the generic pushing system is able to accomplish a given task using the reasonable parameter settings described in Section 2.

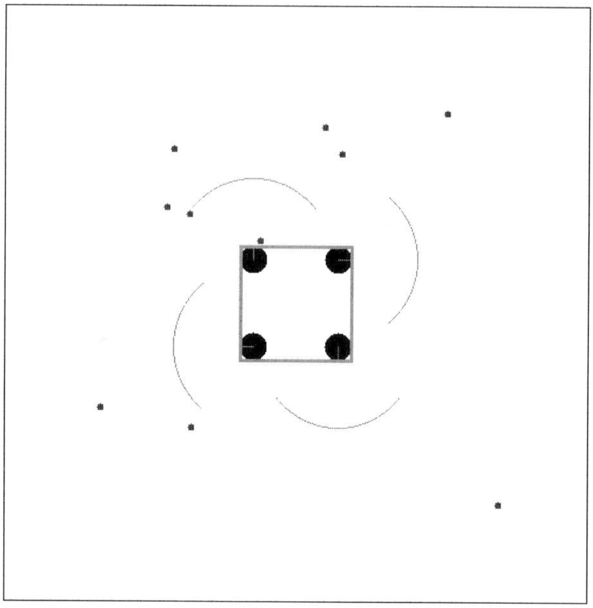

Figure 3. An initial configuration for the foraging task where the four big circles represent robots, the small dots represent objects to be foraged, and the square indicates the home. The arc in front of each robot indicates the angular and linear ranges of each robot's sensor.

3.1 Foraging Proof of Concept

The environment was set up with 10 randomly placed attractors in a 10x10 unit field. The attractors were placed according to a uniform distribution, rejecting (unpushable) points within 0.5 units of a wall, and positions that lie within the home zone. Four robots began on the edges of the home zone, which is centered at the point (0,0). The initial configuration of the robots is shown in Figure 3, with their home zone roughly outlined with the center square. All variables were held constant and set to $c = 0$, $p = 1$, $b = false$.

For each randomly generated environment configuration, 10 trials were run with each controller, and the results were recorded in terms of both completion time (measured in simulated milliseconds), and in terms of the number of attractors successfully foraged, called the success rate from this point on. The simulation was set to timeout after 2,620 simulator seconds, which corresponded to roughly 3 minutes of real-time when watching the simulator live. Any incomplete runs were simply treated as having taken 2,620 simulator seconds. 30 maps were generated and tested. We will report the average completion time and success rate for each controller.

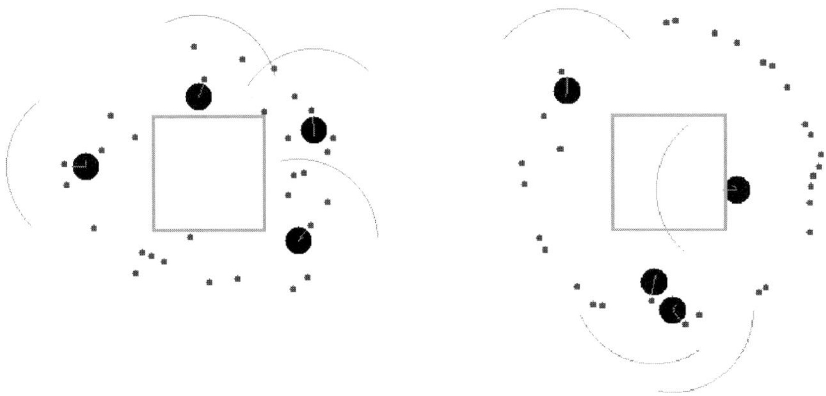

Figure 4. An intermediate (left) and final (right) configurations for the construction task of a circular nest site. The graphic symbols have the same interpretations as those in Figure 3. Notice the approximately circular structure that is created near the end of the task on the right.

3.2 Construction Proof of Concept

The purpose of this experiment is to determine whether or not the generic pusher, using reasonable parameter values, can consistently complete the nest construction task. The environment was set up with 30 randomly placed attractors in a 10x10 field (shown in Figure 4). Home was defined as the point (0,0), and construction was considered successful when all attractors were at least 3 units from home. Attractors were placed in a uniform fashion, rejecting any points generated less than 1 unit from home, or more than 3 units from home. Four robots were placed near home, in the initial configuration shown in Figure 4. The robots were controlled using the generic pusher with $c = 0$, $p = 1$, $b = true$. Three random initial attractor placements were generated, and 10 trials were run for each case. Completion time and success rate were recorded.

3.3 Box-Pushing Proof of Concept

The environment was set up with one circular attractor of radius 1 located at (-2,2.5). The objective was to push the attractor so its center point lies within 1 unit of home at (0,0). Between 1 and 4 robots were arrayed around the attractor, in the configuration shown in Figure 5. A robot with label x indicates that it

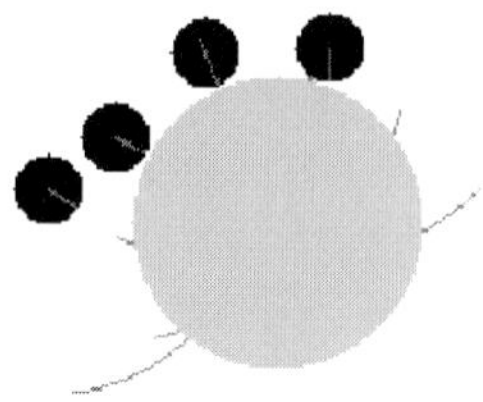

Figure 5. A group transport experiment in progress with four robots and one large object to be moved.

is present so long as at least x robots are present. One will note that all four robots can immediately see the attractor. This removes the search component from the task, so we are observing only whether or not they are capable of pushing the box with the given p setting.

Three variables are manipulated during this experiment. The attractor is given a weight w drawn from $\{0,1,2\}$. The position parameter is drawn from $\{0,1,2,3,4\}$. Finally, the number of robots present varies from 1 to 4. All combinations were tested, for a total of 45 valid configurations (not counting those configurations that did not have enough robots to possibly push the attractor), with 10 trials conducted on each configuration. As above, the robots were given 2,620 simulation second to complete the task. The average success rate was stored for each configuration. We do not concern ourselves with completion time for this experiment, as the focus is on whether or not the task is possible at all given the configuration.

3.4 Results and Discussion

The results of the construction proof of concept test were very encouraging. In all trials conducted on three randomly generated maps, the team of four robots was able to accomplish the construction task in less than 505 simulation seconds, which is well bellow our time limit of 2,620 for other tasks. The results were pleasing on a qualitative level as well, often producing an even ring of attractors around home.

Regarding the results for the foraging task, the maximum success rate achievable was to forage all (100%) possible attractors. With cooperation $c = 0$ and $c = 1$, the success rate did not vary much and was at 95% and 94%, respec-

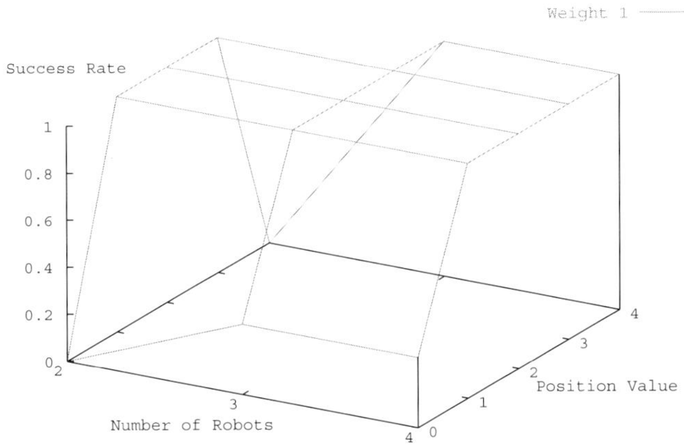

Figure 6. Success rates by parameter value and robot count for collective transport of an attractor with weight 1.

tively. However, the average completion time with $c = 0$ was significantly lower at 1,297 than with $c = 1$ at 1,568, both in simulation steps.

For the group transport task, for any object weight, it was always possible to achieve a 100% success rate, provided that the position parameter p was chosen properly and that a sufficient number of robots participated in the task. As an example, Figure 6 shows success rate when teams of 2 to 4 robots attempt to push a weight 1 attractor, which requires at least 2 robots to move. We can see that the success rate becomes more sensitive to the p parameter. No teams achieve 100% success rate at $p = 0$, and at $p = 4$ teams of 2 fail completely. The moderate p values within $[1, 3]$ remain very safe, though.

4. Conclusions and Future Work

In this paper, we have attempted to establish the argument that environment is indeed very much responsible for the external behavior exhibited by a group of robots. Our supporting evidence is the successful execution of three collective robotics tasks that employ essentially the same generic behavior-based controller. The controller uses pushing as its universal method for object movement, to perform foraging, group transport, and collective construction, with only trivial adjustment of three parameters that reflect the weights or polarity of the internal robot behaviors.

Through simulated experiments, conducted with the TeamBots simulator, we have shown foraging to be robust to the cooperation parameter c with respect to success rate. In addition, we have found the system to be quite sensitive to the position parameter p when attempting group transport. Though parameter values exist that ensure consistent task completion, it remains the user's burden to find and set those parameters according to object weight and team size. Construction task presented the least challenge and was always completed successful under a wide range of system parameters.

In the future, we would like to investigate the more general form of the problem by considering obstacle avoidance. In such a system, a pushing robot is actually drawn toward obstacles in order to place itself between the obstacle and the attractor, and therefore push the object away from the obstacle. This may allow for the inclusion of static obstacles. We would also like to test whether or not lowering the weight given to obstacle avoidance when cooperating to push an object would reduce the system's sensitivity to the p parameter. A general investigation into stagnation recovery techniques to eliminate p-sensitivity from the system would also be quite useful and informative. Finally, it will be an interesting challenge to study the possibility for the robots to identify a task domain through sensing and adjust their operating parameters automatically.

References

Arkin, R. (1998). *Behavior-Based Robotics*. The MIT Press.

Balch, T. and Arkin, R. (1995). Communication in reactive multiagent robotic systems. *Autonomous Robots*, 1(1):27–52.

Balch, T. and Parker, L. E. (2002). *Robot Teams*. A K Peters.

Brooks, R. (1986). A robust layered control system for a mobile robot. *IEEE Journal of Robotics and Automation*, 2(1):14–23.

Gat, E. (1997). *On three-layer architectures*, pages 195–210. MIT/AAAI.

Kube, C. and Zhang, H. (1993). Collective robotics: From social insects to robots. *Adaptive Behavior*, 2(2):189–219.

Kube, C. and Zhang, H. (1997). Task modelling in collective robotics. *Autonomous Robots*, 4(1):53–72.

Mataric, M. J. (1995). Designing and understanding adaptive group behavior. *Adaptive Behavior*, 4(1):51–80.

Parker, C., Zhang, H., and Kube, R. (2003). Blind bulldozing: Multiple robot nest construction. In *IROS2003*, Las Vegas.

III

INFORMATION / SENSOR SHARING AND FUSION

DECENTRALIZED COMMUNICATION STRATEGIES FOR COORDINATED MULTI-AGENT POLICIES

Maayan Roth, Reid Simmons, and Manuela Veloso
Robotics Institute
Carnegie Mellon University
Pittsburgh, PA 15213
mroth@andrew.cmu.edu, reids@cs.cmu.edu, veloso@cs.cmu.edu

Abstract Although the presence of free communication reduces the complexity of multi-agent POMDPs to that of single-agent POMDPs, in practice, communication is not free and reducing the amount of communication is often desirable. We present a novel approach for using centralized "single-agent" policies in decentralized multi-agent systems by maintaining and reasoning over the possible *joint beliefs* of the team. We describe how communication is used to integrate local observations into the team belief as needed to improve performance. We show both experimentally and through a detailed example how our approach reduces communication while improving the performance of distributed execution.[1]

Keywords: Communication, distributed execution, decentralized POMDP

1. Introduction

Multi-agent systems and multi-robot teams can be used to perform tasks that could not be accomplished by, or would be very difficult with, single agents. Such teams provide additional functionality and robustness over single-agent systems, but also create additional challenges. In any physical system, robots must reason over, and act under, uncertainty about the state of the environment. However, in many multi-agent systems there is additional uncertainty about the collective state of the team. If the agents can maintain sufficient collective belief about the state of the world, they can coordinate their joint actions to achieve high reward. Conversely, uncoordinated actions may be costly.

Just as Partially Observable Markov Decision Problems (POMDPs) are used to reason about uncertainty in single-agent systems, there has been recent interest in using multi-agent POMDPs for coordination of teams of agents (Xuan and Lesser, 2002). Unfortunately, multi-agent POMDPs are known

L.E. Parker et al. (eds.),
Multi-Robot Systems. From Swarms to Intelligent Automata. Volume III, 93–105.

to be highly intractable (Bernstein et al., 2000). Communicating (at zero cost) at every time step reduces the computational complexity of a multi-agent POMDP to that of a single agent (Pynadath and Tambe, 2002). Although single-agent POMDPs are also computationally challenging, a significant body of research exists that addresses the problem of efficiently finding near-optimal POMDP policies (Kaelbling et al., 1998). However, communication is generally not free, and forcing agents to communicate at every time step wastes a limited resource.

In this paper, we introduce an approach that exploits the computational complexity benefits of free communication at policy-generation time, while at run-time maintains agent coordination and chooses to communicate only when there is a perceived benefit to team performance. Section 2 of this paper gives an overview of the multi-agent POMDP framework and discusses related work. Sections 3 and 5 introduce our algorithm for reducing the use of communication resources while maintaining team coordination. Section 4 illustrates this algorithm in detail with an example and Section 6 presents experimental results that demonstrate the effectiveness of our approach at acting in coordination while reducing communication.

2. Background and related work

There are several equivalent multi-agent POMDP formulations (i.e. DEC-POMDP (Bernstein et al., 2000), MTDP (Pynadath and Tambe, 2002), POIPSG (Peshkin et al., 2000)). In general, a multi-agent POMDP is an extension of a single-agent POMDP where α agents take individual actions and receive local observations, but accumulate a joint team reward. The multi-agent POMDP model consists of the tuple $\langle \mathcal{S}, \mathcal{A}, \mathcal{T}, \Omega, \mathcal{O}, \mathcal{R}, \gamma \rangle$, where $\mathcal{S}$ is the set of n world states, and $\mathcal{A}$ is the set of m joint actions available to the team, where each joint action, a^i, is comprised of α individual actions $\langle a^i_1 \dots a^i_\alpha \rangle$. Agents are assumed to take actions simultaneously in each time step. The transition function, $\mathcal{T}$, depends on joint actions and gives the probability associated with starting in a particular state s^i and ending in a state s^j after the team has executed the joint action a^k. Although the agents cannot directly observe their current state, s^t, they receive information about the state of the world through Ω, a set of possible joint observations. Each joint observation ω^i is comprised of α individual observations, $\langle \omega^i_1 \dots \omega^i_\alpha \rangle$. The observation function, $\mathcal{O}$, gives the probability of observing a joint observation ω^i after taking action a^k and ending in state s^j. The reward function $\mathcal{R}$ maps a start state and a joint action to a reward. This reward is obtained jointly by all of the agents on the team, and is discounted by the discount factor γ.

Without communication, solving for the optimal policy of a multi-agent POMDP is known to be NEXP-complete (Bernstein et al., 2000), making these problems fundamentally harder than single-agent POMDPs, which are known to be PSPACE-complete (Papadimitriou and Tsitsiklis, 1987). A recent approach presents a dynamic programming algorithm for finding optimal policies for these problems (Hansen et al., 2004). In some domains, dynamic programming may provide a substantial speed-up over brute-force searches, but in general, this method remains computationally intractable. Recent work focuses on finding heuristic solutions that may speed up the computation of locally optimal multi-agent POMDP policies, but these algorithms either place limitations on the types of policies that can be discovered (e.g. limited-memory finite state controllers (Peshkin et al., 2000)), or make strong limiting assumptions about the types of domains that can be solved (e.g. transition-independent systems (Becker et al., 2003)), or may, in the worst case, still have the same complexity as an exhaustive search for the optimal policy (Nair et al., 2003). Another method addresses the problem by approximating the system (in this case, represented as a POIPSG) with a series of smaller Bayesian games (Emery-Montemerlo et al., 2004). This approximation is able to find locally optimal solutions to larger problems than can be solved using exhaustive methods, but is unable to address situations in which a guarantee of strict agent coordination is needed. Additionally, none of these approaches address the issue of communication as a means for improving joint team reward.

Although free communication transforms a multi-agent POMDP into a large single agent POMDP, in the general case where communication is not free, adding communication does not reduce the overall complexity of optimal policy generation for a multi-agent POMDP (Pynadath and Tambe, 2002). Unfortunately, for most systems, communication is not free, and communicating at every time step may be unnecessary and costly. However, it has been shown empirically that adding communication to a multi-agent POMDP may not only improve team performance, but may also shorten the time needed to generate policies (Nair et al., 2004).

In this paper, we introduce an algorithm that takes as input a single-agent POMDP policy, computed as if for a team with free communication, and at run-time, maintains team coordination and chooses to communicate only when it is necessary for improving team performance. This algorithm makes two trade-offs. First, it trades off the need to perform computations at run-time in order to enable the generation of an infinite-horizon policy for the team that would otherwise be highly intractable to compute. Secondly, it conserves communication resources, with the potential trade-off of some amount of reward.

3. Dec-Comm algorithm

Single-agent POMDP policies are mappings from beliefs to actions ($\pi : \mathcal{B} \rightarrow \mathcal{A}$), where a belief, $b \in \mathcal{B}$, is a probability distribution over world states. An individual agent in a multi-agent system cannot calculate this belief because it sees only its own local observations. Even if an agent wished to calculate a belief based only on its own observations, it could not, because the transition and observation functions depend on knowing the joint action of the team.

A multi-agent POMDP can be transformed into a single-agent POMDP by communicating at every time step. A standard POMDP solver can then be used to generate a policy that operates over joint observations and returns joint actions, ignoring the fact that these joint observations and actions are comprised of individual observations and actions. The belief over which this policy operates, which is calculated identically by each member of the team, is henceforth referred to as the *joint belief*.

Creating and executing a policy over joint beliefs is equivalent to creating a centralized controller for the team and requires agents to communicate their observations at each time step. We wish to reduce the use of communication resources. Therefore, we introduce the DEC-COMM algorithm that:

- in a decentralized fashion, selects actions based on the possible joint beliefs of the team
- chooses to communicate when an agent's local observations indicate that sharing information would lead to an increase in expected reward

3.1 Q-POMDP: Reasoning over possible joint beliefs

The Q-MDP method is an approach for finding an approximate solution to a large single-agent POMDP by using the value functions ($\mathcal{V}_a(s)$ is the value of taking action a in state s and henceforth acting optimally) that are easily obtainable for the system's underlying MDP (Littman et al., 1995). In Q-MDP, the best action for a particular belief, b, is chosen according to $\text{Q-MDP}(b) = \arg\max_a \sum_{s \in \mathcal{S}} b(s) \times \mathcal{V}_a(s)$, which averages the values of taking each action in every state, weighted by the likelihood of being in that state as estimated by the belief.

Analogously, we introduce the Q-POMDP method for approximating the best joint action for a multi-agent POMDP by reasoning over the values of the possible joint beliefs in the underlying centralized POMDP. In our approach, a joint policy is created for the system, as described above. During execution, each agent calculates a tree of possible joint beliefs of the team. These joint beliefs represent all of the possible observation histories that could have been observed by the team. We define $\mathcal{L}^t$, the set of leaves of the tree at depth t, to be the set of possible joint beliefs of the team at time t. Each $\mathcal{L}_i^t$ is a tuple consisting of $\langle b^t, p^t, \vec{\omega}^t \rangle$, where $\vec{\omega}^t$ is the joint observation history that would lead to $\mathcal{L}_i^t$, b^t is the joint belief at that observation history, and p^t is the probability of the team observing that history.

Figure 1 presents the algorithm for expanding a single leaf in a tree of possible joint beliefs. Each leaf has a child leaf for every possible joint observation.

For each observation, $Pr(\omega^i|a,b^t)$, the probability of receiving that observation while in belief state b^t and having taken action a, is calculated. The resulting belief, b^{t+1}, is calculated using a standard Bayesian update (Kaelbling et al., 1998). The child leaf is composed of this new belief, b^{t+1}, the probability of reaching that belief, which is equivalent to the probability of receiving this particular observation in the parent leaf times the probability of reaching the parent leaf, and the corresponding observation history. Note that this algorithm entirely ignores the actual observations seen by each agent, enabling the agents to compute identical trees in a decentralized fashion.

GROWTREE($\mathcal{L}_i^t$, a)
  $\mathcal{L}^{t+1} \leftarrow \emptyset$
  for each $\omega^j \in \Omega$
    $b^{t+1} \leftarrow \emptyset$
    $Pr(\omega^j|a,b^t) \leftarrow \sum_{s' \in \mathcal{S}} \mathcal{O}(s',a,\omega^j) \sum_{s \in \mathcal{S}} \mathcal{T}(s,a,s')b^t(s)$
    for each s' $\in \mathcal{S}$
      $b^{t+1}(s') \leftarrow \dfrac{\mathcal{O}(s',a,\omega^j \sum_{s \in \mathcal{S}} \mathcal{T}(s,a,s')b^t(s)}{Pr(\omega^j|a,b^t)}$
    $p^{t+1} \leftarrow p(\mathcal{L}_i^t) \times Pr(\omega^j|a,b^t)$
    $\vec{\omega}^{t+1} \leftarrow \vec{\omega}(\mathcal{L}_i^t) \circ \langle \omega^j \rangle$
    $\mathcal{L}^{t+1} \leftarrow \mathcal{L}^{t+1} \cup [b^{t+1}, p^{t+1}, \vec{\omega}^{t+1}]$
  return $\mathcal{L}^{t+1}$

Figure 1. Algorithm to grow the children of one leaf in a tree of possible beliefs

The Q-POMDP heuristic, Q-POMDP$(\mathcal{L}^t) = \arg\max_a \sum_{\mathcal{L}_i^t \in \mathcal{L}^t} p(\mathcal{L}_i^t) \times \mathcal{Q}(b(\mathcal{L}_i),a)$, selects a single action that maximizes expected reward over all of the possible joint beliefs. Because this reward is a weighted average over several beliefs, there may exist domains for which an action that is strictly dominated in any single belief, and therefore does not appear in the policy, may be the optimal action when there is uncertainty about the belief. We define the $\mathcal{Q}$ function, $\mathcal{Q}(b^t,a) = \sum_{s \in \mathcal{S}} \mathcal{R}(s,a)b^t(s) + \gamma \sum_{\omega \in \Omega} Pr(\omega|a,b^t)\mathcal{V}^{\pi}(b^{t+1})$, in order to take these actions into account. The value function, $\mathcal{V}^{\pi}(b)$, gives the maximum attainable value at the belief b, but is only defined over those actions which appear in the single-agent policy π. The $\mathcal{Q}$ function returns expected reward for any action and belief. b^{t+1} is the belief that results from taking action a in belief state b^t and receiving the joint observation ω. $Pr(\omega|a,b^t)$ and b^{t+1} are calculated as in Figure 1.

Since all of the agents on a team generate identical trees of possible joint beliefs, and because Q-POMDP selects actions based only on this tree, ignoring the actual local observations of the agents, agents are guaranteed to select the same joint action at each time step. However, this joint action is clearly very conservative, as agents are forced to take into account all possible contingencies. The DEC-COMM algorithm utilizes communication to allow agents to integrate their actual observations into the possible joint beliefs, while still maintaining team synchronization.

3.2 Dec-Comm: Using communication to improve performance

An agent using the DEC-COMM algorithm chooses to communicate when it sees that integrating its own observation history into the joint belief would cause a change in the joint action that would be selected. To decide whether or not to communicate, the agent computes a_{NC}, the joint action selected by the Q-POMDP heuristic based on its current tree of possible joint beliefs. It then prunes the tree by removing all beliefs that are inconsistent with its own observation history and computes a_C, the action selected by Q-POMDP based on this pruned tree. If the actions are the same, the agent chooses not to communicate. If the actions are different, this indicates that there is a potential gain in expected reward through communication, and the agent broadcasts its observation history to its teammates. When an agent receives a communication from one of its teammates, it prunes its tree of joint beliefs to be consistent with the observations communicated to it, and recurses to see if this new information would lead it to choose to communicate. Because there may be multiple instances of communication in each time step, agents must wait a fixed period of time for the system to quiesce before acting. Figure 2 provides the details of the DEC-COMM algorithm.

4. Example

To illustrate the details of our algorithm, we present an example in the two-agent tiger domain introduced by Nair *et al.* (Nair et al., 2003). We use the tiger domain because it is easily understood, and also because it is a problem that requires coordinated behavior between the agents. The tiger problem consists of two doors, LEFT and RIGHT. Behind one door is a tiger, and behind the other is a treasure. S consists of two states, SL and SR, indicating respectively that the tiger is behind the left door or the right door. The agents start out with a uniform distribution over these states ($b(\text{SR}) = 0.5$).

Each agent has three individual actions available to it: OPENL, which opens the left door, OPENR, which opens the right door, and LISTEN, an information-gathering action that provides an observation about the location of the tiger. Together, the team may perform any combination of these individual actions. A joint action of $\langle$LISTEN, LISTEN$\rangle$ keeps the world in its current state. In order to make this an infinite-horizon problem, if either agent opens a door, the world is randomly and uniformly reset to a new state. The agents receive

```
DEC-COMM(L^t, ω⃗^t_j)
    a_NC ← Q-POMDP(L^t)
    L' ← prune leafs inconsistent with ω⃗^t_j from L^t
    a_C ← Q-POMDP(L')
    if a_NC ≠ a_C
        communicate ω⃗^t_j to the other agents
        return DEC-COMM(L', ∅)
    else
        if communication ω⃗^t_k was received from another agent k
            L^t ← prune leafs inconsistent with ω⃗^t_k from L^t
            return DEC-COMM(L^t, ω⃗^t_j)
        else
            take action a_NC
            receive observation ω^{t+1}_j
            ω⃗^{t+1}_j ← ω⃗^t_j ∘ ⟨ω^{t+1}_j⟩
            L^{t+1} ← ∅
            for each L^t_i ∈ L^t
                L^{t+1} ← L^{t+1} ∪ GROWTREE(L^t_i, a_NC)
            return [L^{t+1}, ω⃗^{t+1}_j]
```

Figure 2. One time step of the DEC-COMM algorithm for an agent j

two observations, HL and HR, corresponding to hearing the tiger behind the left or right door. For the purposes of our example, we modify the observation function from the one given in Nair *et al.* If a door is opened, the observation is uniformly chosen and provides no information; the probability of an individual agent hearing the correct observation if both agents LISTEN is 0.7. (Observations are independent, so the joint observation function can be computed as the cross-product of the individual observation functions.) This change makes it such that the optimal policy is to hear two consistent observations (e.g. HR, HR) before opening a door.

The reward function for this problem is structured to create an explicit coordination problem between the agents. The highest reward (+20) is achieved when both agents open the same door, and that door does not contain the tiger. A lower reward (-50) is received when both agents open the incorrect door. The worst case is when the agents open opposite doors (-100), or when one agent opens the incorrect door while the other agent listens (-101). The cost of ⟨LISTEN, LISTEN⟩ is -2. We generated a joint policy for this problem with Cassandra's POMDP solver (Cassandra,), using a discount factor of $\gamma = 0.9$. Note that although there are nine possible joint actions, all actions other than ⟨OPENL, OPENL⟩, ⟨OPENR, OPENR⟩, and ⟨LISTEN, LISTEN⟩ are strictly dominated, and we do not need to consider them.

Time Step 0: In this example, the agents start out with a synchronized joint belief of $b(\mathrm{SR}) = 0.5$. According to the policy, the optimal joint action at this belief is ⟨LISTEN, LISTEN⟩. Because their observation histories are empty, there is no need for the agents to communicate.

Time Step 1: The agents execute ⟨LISTEN, LISTEN⟩, and both agents observe HL. Each agent independently executes GROWTREE. Figure 3 shows the tree of possible joint beliefs calculated by each agent. The Q-POMDP heuristic, executed over this tree, determines that the best possible joint action is ⟨LISTEN, LISTEN⟩.

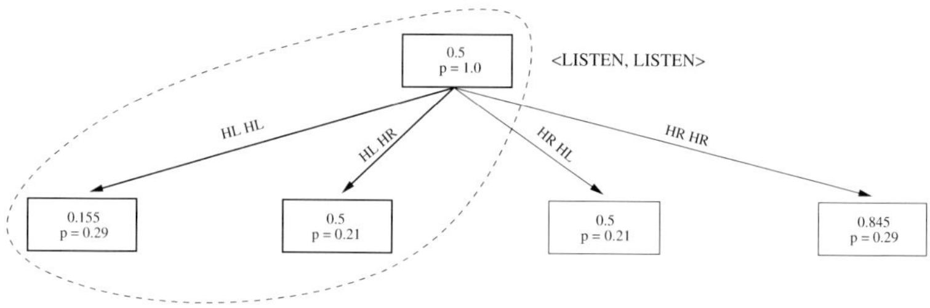

Figure 3. Joint beliefs after a single action

When deciding whether or not to communicate, agent 1 prunes all of the joint beliefs that are not consistent with its having heard HL. The circled nodes in Figure 3 indicate those nodes which are not pruned. Running Q-POMDP on the pruned tree shows that the best joint action is still ⟨LISTEN, LISTEN⟩, so agent 1 decides not to communicate. It is important to note that at this point, a centralized controller would have observed two consistent observations of HL and would perform ⟨OPENR, OPENR⟩. This is an instance in which our algorithm, because it does not yet have sufficient reason to believe that there will be a gain in reward through communication, performs worse than a centralized controller.

Time Step 2: After performing another ⟨LISTEN, LISTEN⟩ action, each agent again observes HL. Figure 4 shows the output of GROWTREE after the second action. The Q-POMDP heuristic again indicates that the best joint action is ⟨LISTEN, LISTEN⟩.

Agent 1 reasons about its communication decision by pruning all of the joint beliefs that are not consistent with its entire observation history (hearing HL twice). This leaves only the nodes that are circled in Figure 4. For the pruned tree, Q-POMDP indicates that the best action is ⟨OPENR, OPENR⟩. Because the pre-communication action, a_{NC}, differs from the action that would be cho-

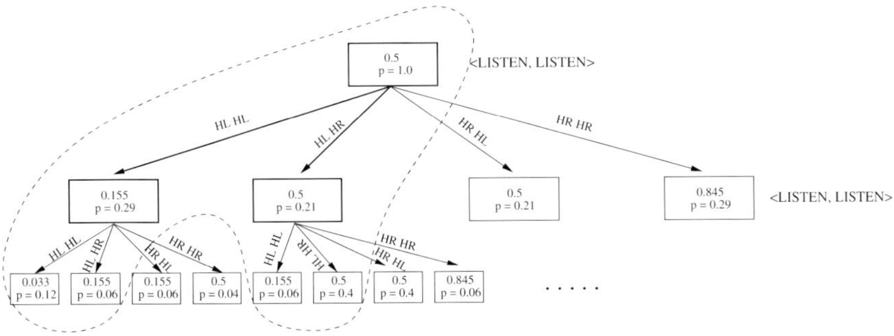

Figure 4. Joint beliefs after the second action

sen post-communication, a_C, agent 1 chooses to communicate its observation history to its teammate.

In the meantime, agent 2 has been performing an identical computation (since it too observed two instances of HL) and also decides to communicate. After both agents communicate, there is only a single possible belief remaining, $b(\text{SR}) = 0.033$. The optimal action for this belief is $\langle$OPENR, OPENR$\rangle$, which is now performed by the agents.

5. Particle filter representation

The above example shows a situation in which both agents decide to communicate their observation histories. It is easy to construct situations in which one agent would choose to communicate but the other agent would not, or examples in which both agents would decide not to communicate, possibly for many time steps (e.g. the agents observe alternating instances of HL and HR). From the figures, it is clear that the tree of possible joint beliefs grows rapidly when communication is not chosen. To address cases where the agents do not communicate for a long period of time, we present a method for modeling the distribution of possible joint beliefs using a particle filter.

A particle filter is a sample-based representation that can be used to encode an arbitrary probability distribution using a fixed amount of memory. In the past, particle filters have been used with single-agent POMDPs (i.e. for state estimation during execution (Poupart et al., 2001)). We draw our inspiration from an approach that finds a policy for a continuous state-space POMDP by maximizing over a distribution of possible belief states, represented by a particle filter (Thrun, 2000).

In our approach, each particle, $\mathcal{L}^i$ is a tuple of α observation histories, $\langle \vec{\omega}_a \ldots \vec{\omega}_\alpha \rangle$, corresponding to a possible observation history for each agent. Taken together, these form a possible joint observation history, and along with the system's starting belief state, b^0, and the history of joint actions taken by the team, $\vec{a}$, uniquely identify a possible joint belief. Every agent stores two particle filters, $\mathcal{L}_{joint}$, which represents the joint possible beliefs of the team, pruned only by communication, and $\mathcal{L}_{own}$, those beliefs that are consistent with the agent's own observation history. Belief propagation is performed for these filters as described in (Thrun, 2000), with the possible next observations for $\mathcal{L}_{joint}$ taken from all possible joint observations, and the possible next observations for $\mathcal{L}_{own}$ taken only from those joint observations consistent with the agent's own local observation at that time step.

The DEC-COMM algorithm proceeds as described in Section 3, with $\mathcal{L}_{joint}$ used to generate a_{NC} and $\mathcal{L}_{own}$ used to generate a_c. The only complication arises when it comes time to prune the particle filters as a result of communication. Unlike the tree described earlier that represents the distribution of possible joint beliefs exactly, a particle filter only approximates the distribution. Simply removing those particles not consistent with the communicated observation history and resampling (to keep the total number of particles constant) may result in a significant loss of information about the possible observation histories of agents that have not yet communicated.

Looking at the example presented in Section 4, it is easy to see that there is a correlation between the observation histories of the different agents. (i.e. If one agent observes $\langle \text{HL}, \text{HL} \rangle$, it is unlikely that the other agent will have observed $\langle \text{HR}, \text{HR} \rangle$.) To capture this correlation when pruning, we define a similarity metric between two observation histories, Figure 5. When an observation history $\vec{\omega}_i^t$ has been communicated by agent i, to resample the new $\mathcal{L}_{joint}$, the observation history in each particle corresponding to agent i is compared to $\vec{\omega}_i^t$. The comparison asks the question, "Suppose an agent has observed $\vec{\omega}_i^t$ after starting in belief b^0 and knowing that the team has taken the joint action history $\vec{a}^t$. What is the likelihood that an identical agent would have observed the observation history $\vec{\omega}_j^t$?" The value returned by this comparison is used as a weight for the particle. The particles are then resampled according to the calculated weights, and the agent i observation history for each particle is replaced with $\vec{\omega}_i^t$.

6. Results and analysis

We demonstrate the performance of our approach experimentally by comparing the reward achieved by a team that communicates at every time step (i.e. a centralized controller) to a team that uses the DEC-COMM algorithm to select actions and make communication decisions. We ran our experiment on the two-agent tiger domain as described in Section 4. In each experiment, the world state was initialized randomly, and the agents were allowed to act for 8 time steps. The team using a particle representation used 2000 samples to

SIMILARITY($\vec{\omega}_i^t, \vec{\omega}_j^t, \vec{a}^t$)
    $sim \leftarrow 1$
    $b \leftarrow b^0$
    for $t' = 1 \ldots t$
        for each $s \in \mathcal{S}$
            $b(s) \leftarrow \mathcal{O}(s, a^{t'}, \omega_i^{t'}) b(s)$
        normalize b
        $sim \leftarrow sim \times \sum_{s \in \mathcal{S}} \mathcal{O}(s, a^{t'}, \omega_j^{t'}) b(s)$
        for each $s \in \mathcal{S}$
            $b(s) \leftarrow \sum_{s' \in \mathcal{S}} \mathcal{T}(s', a^{t'}, s) b(s)$
        normalize b
    return sim

Figure 5. The heuristic used to determine the similarity between two observation histories, where $\vec{\omega}_i^t$ is the true (observed) history

represent the possible beliefs. We ran 30000 trials of this experiment. Table 1 summarizes the results of these trials.

Table 1. Experimental results

	μ_{Reward}	σ_{Reward}	μ_{Comm}	σ_{Comm}
Full Comm.	17.0	37.9	16	0
DEC-COMM (tree)	8.9	28.9	2.9	1.1
DEC-COMM (particles)	9.4	30.3	2.6	1.0

It may appear at first glance as though the performance of the DEC-COMM algorithm is substantially worse than the centralized controller. However, as the high standard deviations indicate, the performance of even the centralized controller varies widely, and DEC-COMM under-performs the fully communicating system by far less than one standard deviation. Additionally, it achieves this performance by using less than a fifth as much communication as the fully communicating system. Note that the particle representation performs comparably to the tree representation (within the error margins), indicating that with a sufficient number of particles, there is no substantial loss of information.

We are currently working on comparing the performance of our approach to COMMUNICATIVE JESP, a recent approach that also uses communication to improve the computational tractability and performance of multi-agent POMDPs (Nair et al., 2004). However, this comparison is difficult for several reasons. First of all, the COMMUNICATIVE JESP approach treats communication as domain-level action in the policy. Thus, if an agent chooses to communicate in a particular time step, it cannot take an action. More significantly, their approach deals only with synchronized communications, meaning that if one agent on a team chooses to communicate, it also forces all its other teammates to communicate at that time step.

7. Conclusion

We present in this paper an approach that enables the application of centralized POMDP policies to distributed multi-agent systems. We introduce the novel concept of maintaining a tree of possible joint beliefs of the team, and describe a heuristic, Q-POMDP, that allows agents to select the best action over the possible beliefs in a decentralized fashion. We show both through a detailed example and experimentally that our DEC-COMM algorithm makes communication decisions that improve team performance while reducing the instances of communication. We also provide a fixed-size method for maintaining a distribution over possible joint team beliefs.

In the future, we are interested in looking at factored representations that may reveal structural relationships between state variables, allowing us to address the question of *what* to communicate, as well was *when* to communicate. Other areas for future work include reasoning about communicating only *part* of the observation history, and exploring the possibility of agents *asking* their teammates for information instead of only *telling* what they know.

Notes

1. This work has been supported by several grants, including NASA NCC2-1243, and by Rockwell Scientific Co., LLC under subcontract no. B4U528968 and prime contract no. W911W6-04-C-0058 with the US Army. This material was based upon work supported under a National Science Foundation Graduate Research Fellowship. The views and conclusions contained in this document are those of the authors and should not be interpreted as necessarily representing the official policies or endorsements, either expressed or implied, by the sponsoring institutions, the U.S. Government or any other entity.

References

Becker, R., Zilberstein, S., Lesser, V., and Goldman, C. V. (2003). Transition-independent decentralized Markov Decision Processes. In *International Joint Conference on Autonomous Agents and Multi-agent Systems*.

Bernstein, D. S., Zilberstein, S., and Immerman, N. (2000). The complexity of decentralized control of Markov Decision Processes. In *Uncertainty in Artificial Intelligence*.

Cassandra, A. R. POMDP solver software. http://www.cassandra.org/pomdp/code/index.shtml.

Emery-Montemerlo, R., Gordon, G., Schneider, J., and Thrun, S. (2004). Approximate solutions for partially observable stochastic games with common payoffs. In *International Joint Conference on Autonomous Agents and Multi-Agent Systems*.

Hansen, E. A., Bernstein, D. S., and Zilberstein, S. (2004). Dynamic programming for partially observable stochastic games. In *National Conference on Artificial Intelligence*.

Kaelbling, L. P., Littman, M. L., and Cassandra, A. R. (1998). Planning and acting in partially observable domains. *Artificial Intelligence*.

Littman, M. L., Cassandra, A. R., and Kaelbling, L. P. (1995). Learning policies for partially observable environments: Scaling up. In *International Conference on Machine Learning*.

Nair, R., Pynadath, D., Yokoo, M., Tambe, M., and Marsella, S. (2003). Taming decentralized POMDPs: Towards efficient policy computation for multiagent settings. In *International Joint Conference on Artificial Intelligence*.

Nair, R., Roth, M., Yokoo, M., and Tambe, M. (2004). Communication for improving policy computation in distributed POMDPs. In *International Joint Conference on Autonomous Agents and Multi-agent Systems*.

Papadimitriou, C. H. and Tsitsiklis, J. N. (1987). The complexity of Markov Decision Processes. *Mathematics of Operations Research*.

Peshkin, L., Kim, K.-E., Meuleau, N., and Kaelbling, L. P. (2000). Learning to cooperate via policy search. In *Uncertainty in Artificial Intelligence*.

Poupart, P., Ortiz, L. E., and Boutilier, C. (2001). Value-directed sampling methods for monitoring pomdps. In *Uncertainty in Artificial Intelligence*.

Pynadath, D. V. and Tambe, M. (2002). The communicative Multiagent Team Decision Problem: Analyzing teamwork theories and models. *Journal of AI Research*.

Thrun, S. (2000). Monte carlo pomdps. In *Neural Information Processing Systems*.

Xuan, P. and Lesser, V. (2002). Multi-agent policies: From centralized ones to decentralized ones. In *International Joint Conference on Autonomous Agents and Multi-agent Systems*.

IMPROVING MULTIROBOT MULTITARGET TRACKING BY COMMUNICATING NEGATIVE INFORMATION

Matthew Powers, Ramprasad Ravichandran, Frank Dellaert, Tucker Balch
Borg Lab
College of Computing
Georgia Institute of Technology
Atlanta, Georgia 30332–0250
{mpowers, raam, dellaert, tucker} @cc.gatech.edu

Abstract In this paper, we consider the sensor fusion problem for a team of robots, each equipped with monocular color cameras, cooperatively tracking multiple ambiguous targets. In addition to coping with sensor noise, the robots are unable to cover the entire environment with their sensors and may be out numbered by the targets. We show that by explicitly communicating *negative information* (i.e. where robots *don't* see targets), tracking error can be reduced significantly in most instances. We compare our system to a baseline system and report results.

Keywords: multirobot, multitarget tracking, sensor fusion, negative information

1. Introduction

The problem of using multiple robots to track multiple targets has been approached from several angles. Some previous work (Parker, 1997),(Parker, 1999),(Werger and Matarić, 2000) deals with the problem of allocating robotic resources to best observe the targets, while other work (Reid, 1979),(Schulz and Cremers, 2001),(Khan and Dellaert, 2003a) deals with probabilistically tracking multiple targets from a single or static vantage point. In this work, we deal with the sensor fusion problem for multiple moving observer robots cooperatively tracking multiple ambiguous moving targets. It is assumed the robots have a limited sensor range and the robots' mission is not exclusively to track the targets, but to keep track of the targets while performing other tasks (which may or may not require accurate knowledge of the targets' positions). Due to this final constraint, we do not assume we may move the robots for the

L.E. Parker et al. (eds.),
Multi-Robot Systems. From Swarms to Intelligent Automata. Volume III, 107–117.

purpose of sensing, but we must make the most effective use of the information we have. It is likely the observing robots are unable to see all the targets simultaneously, individually or collectively.

This scenario is motivated by, although not unique to, the problem in robot soccer (http://www.robocup.org/) of keeping track of one's opponents. In the opponent tracking problem, a team of robots must maintain a belief about the position of a team of identical opponent robots in an enclosed field. The observing and target (opponent) robots are constantly moving and performing several tasks in parallel, such as chasing the ball and localization. Observing robots are usually unable to act in a way to optimize their observations of their opponents since acting with respect to the ball is the primary objective. While the observing robots may not be able to act with respect to their opponents, it is still advantageous to accurately estimate their opponents' positions since that information can be used to improve the value of their actions on the ball (e.g. passing the ball to a teammate not covered by an opponent).

We address this problem by communicating a relatively small amount of information among a team of robots, fusing observations of multiple targets using multiple particle filters. Importantly, *negative information* is communicated, along with observations of targets, in the form of parameters to a sensor model. When the robots are not able to see all the targets simultaneously, negative information allows the robots to infer information about the unobserved targets. While each robot's sensor model is assumed to be identical in our experiments, heterogeneous sensor models could easily be used, allowing heterogeneous robot teams to function in the same way.

Our system has been tested in laboratory conditions, using moving observer robots and targets. The data gathered was analyzed offline so that other methods could be tested on the same data. Noting prior art (Schulz and Cremers, 2001), we expect that the algorithms described in this paper can be efficiently run on board the robot platforms used to gather data in our experiments. It is also expected that the results of this paper will be applicable to a wide range of multirobot, multitarget tracking problems.

2. Problem Definition

This work is concerned with the sensor fusion problem of tracking multiple moving targets with a cooperative multirobot system. It is assumed that the robots cannot act with respect to the targets (e.g. move so as to optimize their view of the targets).

More formally, a team of m robots R must track a set of n targets O within an enclosed space S. The members of R move independently of the members of O and vice versa. R's sensors may or may not be able to cover the entire space S or observe all the members of O at a given timestep t. At each timestep

t, each robot $r_i \in R$ produces a measurement, (which may be null), z_t^{ij}, of each target $o_j \in O$.

The robots may communicate information with their teammates to collectively estimate the position of each target at each timestep. The goal of the team is to minimize error from ground truth of each position estimate.

3. Related Work

3.1 Multirobot Sensing

In (Parker, 1997) Parker defines the problem of Cooperative Multirobot Observation of Multiple Moving Targets (CMOMMT) as follows. Given: S, a bounded enclosed two-dimensional region; R, a team of team of m robots; $O(t)$, a set of n targets such that the binary function $In(o_j(t), S)$ returns *true* when target $o_j \in O(t)$ is within the region S at time t; $A(t)$ an $m \times n$ matrix defined so

$$a_{ij}(t) = \begin{cases} 1 & \text{if robot } \mathrm{r_i} \text{ is monitoring target } o_j(t) \text{ in S at time t} \\ 0 & \text{otherwise} \end{cases} \tag{1}$$

and the *logical OR* operator defined as

$$\bigvee_{i=1}^{k} h_i = \begin{cases} 1 & \text{if there exists an } i \text{ such that } h_i = 1 \\ 0 & \text{otherwise} \end{cases} \tag{2}$$

the goal of CMOMMT is to maximize the value

$$\frac{\sum_{t=0}^{T} \sum_{j=1}^{m} \bigvee_{i=1}^{k} a_{ij}(t)}{T \times m} \tag{3}$$

This problem has been approached in several ways, with each approach showing its own strengths. In (Parker, 1997),(Parker, 1999) the ALLIANCE architecture was used to allocate robotic resources for the CMOMMT task. Task allocation was done implicitly using one-way communication. In contrast, Werger and Matarić (Werger and Matarić, 2000) used explicit communication to assign robots to targets.

While the above systems worked within relatively simple open environments, Jung and Sukhatme (Jung and Sukhatme, 2002b),(Jung and Sukhatme, 2002c),(Jung and Sukhatme, 2002a) worked in more complex environments. Their approach was to deploy robots to different regions of the environment using a topological map and estimates of the target density in each region. They show that the region-based approach works well when the target density is high or the environment is subject to occlusions.

Varying somewhat from above stated goal of CMOMMT, Stroupe and Balch proposed MVERT (Stroupe and Balch, 2004a) as a distributed reactive system to navigate robot teams so as to maximize the team's knowledge of the environment. They demonstrate a team of robots moving to gain information about static targets, and then expand (Stroupe, 2003) (Stroupe and Balch, 2004b) to demonstrate a team navigating to optimize its knowledge about a group of moving targets.

Howard et al (Howard and Sukhatme, 2003) present a method for improving localization of a team of robots by using the robots to observe each other and maintain an ego-centric model of the locations of the other members of the robot team.

3.2 Particle filters

Particle filters have been used extensively in the tracking community to represent arbitrary likelihood distributions over multidimensional space (Gordon and Smith, 1993),(Isard and Blake, 1996),(Carpenter and Fernhead, 1997), (Dellaert and Thrun, 1999). Particle filter theory is well-documented; (Arulampalam and Clapp, 2002) and (Khan and Dellaert, 2003a) explain it well. We give a brief review, based on (Khan and Dellaert, 2003a), to aid the understanding of our system.

A particle filter is a Bayesian filter, meaning that the posterior distribution is recursively updated over the current state X_t, (the targets' locations) given all observations $Z_t = \{Z_1, ..., Z_t\}$ up to time t (all the robots' senor readings up to the current time).

$$P(X_t|Z_t) = kP(Z_t|X_t)P(X_t|Z_{t-1}) \tag{4}$$

$$= kP(Z_t|X_t)\int_{X_{t-1}} P(X_t|X_{t-1})P(X_{t-1}|Z_{t-1}) \tag{5}$$

The likelihood $P(Z_t|X_t)$ is known as the *sensor* or *measurement model* and $P(X_t|X_{t-1})$ the *motion model*. k is a normalization factor.

Particle filters approximate the posterior $P(X_{t-1}|Z_{t-1})$ recursively as a set of N samples, or particles, $\{X_{t-1}^{(r)}, \pi_{t-1}^{(r)}\}_{r=1}^{N}$. $\pi_{t-1}^{(r)}$ is the weight for particle $X_{t-1}^{(r)}$. A Monte Carlo approximation of the integral yields:

$$P(X_t|Z_t) \approx kP(Z_t|X_t)\sum_r \pi_{t-1}^{(r)} P(X_t|X_{(t-1)}^{(r)}) \tag{6}$$

Intuitively, the filter functions recursively in two steps:

Prediction – Each particle is moved from its current location according to a stochastic motion model.

Update – Each particle is weighted according to a sensor model. Particles are resampled with replacement from the weighted set. While the number of particles is maintained in the resampled set, particles that were weighted heavily are likely to be resampled multiple times, while particles that were not weighted heavily are likely to be not chosen at all.

3.3 Multitarget tracking

In classical multitarget tracking, targets are kept distinct by performing a data association step after the measurement step. Trackers of this type include the multiple hypothesis tracker (Reid, 1979), and the joint probabilistic data association filter (JPDAF) (Bar-Shalom and Scheffe, 1980),(Fortmann and Scheffe, 1983). These data association techniques have even been adapted to work with particle filters, as in the particle filtering version of JPDAF (Schulz and Cremers, 2001).

When a tracking problem is likely to result in arbitrary or multimodal likelihood distributions, the standard particle filter can be adapted to fit multitarget tracking (Khan and Dellaert, 2003a). The joint particle filter treats each target as an increase in dimensionality of a single filter. Each particle represents a proposed state for all targets being tracked. That is, if ten targets are being tracked across the x-y plane, each particle represents a hypothesis over twenty dimensions, $(x_t^1, y_t^1, x_t^2, y_t^2, \ldots, x_t^{10}, y_t^{10})$. While this method is algorithmically simple, the number of samples needed to represent high dimensional spaces can become intractably high.

In instances where the states represented by a joint particle filter can be divided into several nearly independent subsets, the joint particle filter can be approximated by splitting it up into several parallel filters to reduce the state space of any given filter. In the above example, the twenty dimensions representing the ten targets being tracked over the x-y plane can be divided up into ten nearly independent subsets, $(x_1, y_1), (x_2, y_2), \ldots, (x_{10}, y_{10})$. Each of these subsets can be treated as a separate tracking problem. If the targets are known to interact, Khan et al (Khan and Dellaert, 2003a),(Khan and Dellaert, 2003b) demonstrate several approaches for tracking interacting targets.

4. Approach

4.1 Negative Information

Our approach to minimizing the collective target position estimate error by the multirobot system described in Section 2 is relatively simple, and may be applicable to other multirobot tracking systems. A separate particle filter is used to track each target's position, X_t. Each robot's observations, Z_t^i are broadcast to its teammates. Observations are greedily associated with targets by comparing to the most recent estimates of the targets' positions, X_{t-1}. Solving analytically,

$$P(\{x_t^j\} | \{r_t^i, Z_t^i\}) \propto \prod_i L(\{x_t^j\}; r_t^i, Z_t^i) \times P(\{x_t^j\} | \{r_t^i\}) \tag{7}$$

Since O moves independently of R, and we assume that the members of O move independently of each other,

$$P(\{x_t^j\}|\{r_t^i, Z_t^i\}) \propto \prod_i \prod_j L(x_t^j; r_t^i, z_t^{ij}) \times P(\{x_t^j\}) \tag{8}$$

The key insight of this work is that even if target j is unobservable by robot i, (i.e. $z_t^{ij} = \phi$), the information represented by z_t^{ij} is still usable for tracking the targets. Since it is known that n targets are in the environment, any null observation, $z_t^{ij} = \phi$, can be treated as a negative observation for target j. Accounting for the special case of a negative observation,

$$L(x_t^j; z_t^{ij}, r_t^i) = \begin{cases} L^-(x_t^j; r_t^i) & \text{if } z_t^{ij} = \phi \\ L^+(x_t^j; z_t^{ij}, r_t^i) & \text{otherwise} \end{cases} \tag{9}$$

Because the likelihood of not seeing target j is not uniform across the environment, this *negative information* still adds information to the estimate of the target's position.

4.2 Sensor Models

Developing appropriate sensor models is key to making use of negative information. Because sensor models necessarily vary from platform to platform, every implementation will be unique.

We assume that each measurement Z_t^i consists of n sensor readings (corresponding to the n targets), consisting either of a range and bearing, or, in the case that a target is unobservable by r_i, a null reading:

$$Z_t^i \in \mathcal{M}^n \tag{10}$$

where

$$\mathcal{M} = \phi \cup (\mathcal{R} \times \mathcal{SO}_1) \tag{11}$$

In the case that target j is observable by robot i,

$$z_t^{ij} = \{r, \theta\} \quad x_t^j = \{x_o, y_o\} \quad r_t^i = \{x_r, y_r, \theta_r\} \tag{12}$$

$$r_o = \sqrt{(x_o - x_r)^2 + (y_o - y_r)^2} \tag{13}$$

$$\theta_o = atan2((y_o - y_r), (x_o - x_r)) - \theta_r \tag{14}$$

$$L^+(z_t^{ij}|x_t^j, r_t^i) \propto e^{-(r-r_o)/(2\sigma_r^2)} \times e^{-(\theta-\theta_o)/(2\sigma_\theta^2)} \tag{15}$$

In the case of a null reading, the sensor model reflects the likelihood of not observing target j. Here, the robots are assumed to have a maximum sensor pan angle θ_{max} (i.e. the robots have a limited field of view, and can not see behind

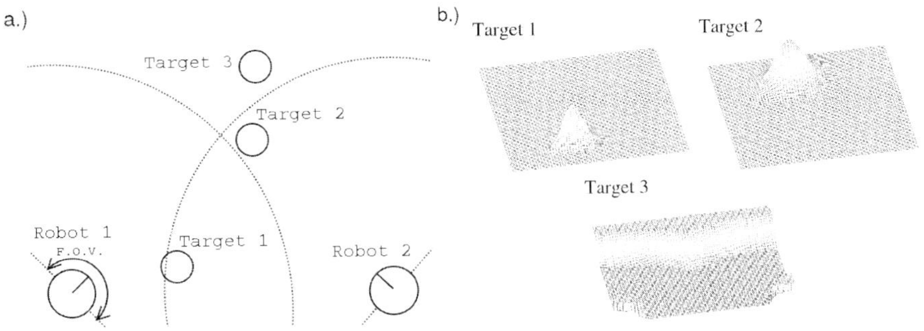

Figure 1. An example of sensor models representing positive and negative information. Figure 1a shows the locations of two robots and three targets Robot 1 can see target 1, while robot 2 can see target 1 and 2. Neither can see target 3. Figure 1b shows the likelihood functions for the location of each target given the observations of the robots.

them). Additionally, it is assumed the robots detect the targets perfectly within the nominal sensor range of r_{nom}, never detect targets beyond the maximum range of r_{max} and detect targets with linearly degrading reliability between r_{nom} and r_{max}.

$$L^{-}(z_t^{ij} = \phi | x_t^j, r_t^i) \propto \begin{cases} 1 & \text{if } \theta_o > \theta_{max} \text{ or } r_o > r_{max} \\ 0 & \text{if } r_o < r_{nom} \\ \frac{r_{max} - r_o}{r_{max} - r_{nom}} & \text{otherwise} \end{cases} \quad (16)$$

Figure 1 give a graphical representation of the above sensor models within the context of two robots tracking three targets. Figure 1a shows the location of the robots and targets in the environment. From its position, robot 1 can observe target 1. Robot 2 can observe targets 1 and 2. Neither robot can observe target 3. Figure 1b shows the respective likelihood functions for the location of each target given the observations of robot 1 and 2 and the combined observations. Note that even though target 3 is not directly observed, some information can be inferred simply from the lack of an observation and the known position of each observing robot.

5. Experimental Approach

5.1 Implementation

Experiments were conducted on a physical robot team of four Sony AIBO robots. The robots used a hybrid software implementation. The filters were implemented in Matlab on a laboratory computer. Low level image processing and motions were implemented in C++ on the robots. CMVision was used for image processing (Bruce and Veloso, 2000). Motion commands were sent from Matlab to the robots and odometry and observation feedback were sent back from the robots to Matlab. All communication was implemented using

TCP/IP over wireless Ethernet. A pair of calibrated overhead cameras and a template-matching algorithm were used to establish ground truth.

The following measured values were used in the robots' sensor model: $\theta_{max} = 110°$, $r_{nom} = 1000mm$, $r_{max} = 1300mm$, $\sigma_r = 150mm$, $\sigma_\theta = 6°$.

5.2 Targets and Landmarks

Six bicolor landmarks of a known size (10 centimeter radius and 20 centimeter height) were used by the robots for localization. The four targets tracked were also bicolor cylinders but had a radius of 8 centimeters and height of 16 centimeters. Although the targets were uniquely colored, this fact was only used for measuring ground truth. The observer robots treated all targets as identical. The targets were mounted on remote-controlled platforms. The size of the environment was restricted to 2.5 meters by 2.5 meters. Figure 2 shows the experimental setup.

5.3 Experiments

Experiments were run using combinations of 1, 2, 3 and 4 observer robots and 1, 2, 3 and 4 targets. At each logical timestep, the targets and the observer robots performed a random walk, all observing robots recorded their observations and ground truth measurements were taken. The percent of the field observable by the robots varied with robots' poses at every timestep and ranged from more than 90% to less than 20%. All targets were not necessarily visible at a given timestep, creating an inherent loss of information.

Several trials of each class of experiment were run with each combination of target and robot numbers. Data was analyzed off-board to allow direct comparisons of tracking techniques on identical data.

A baseline technique was compared against the experimental system. Working under the hypothesis that making use of negative information will improve the error of the tracked target position, the baseline technique was identical to the experimental system with the exception that it did not take into account the negative information sensor model. Error between the ground truth and each tracker was calculated for each timestep of each experiment.

6. Results

Figures 3a and b show the average tracking error over all experiments in each configuration of 1, 2, 3 and 4 observer robots and targets, respectively, for the baseline and experimental systems. Figure 3c shows the difference in the tracking errors between the baseline and experimental systems. Positive numbers in this graph show a reduction in error in the experimental system over the baseline system.

Figure 2. The experimental setup, including four observer robots and four targets. The grid on the ground was used to calibrate the overhead ground truth camera.

Note that the largest reductions in error occur when the observer robots are most outnumbered by the targets. This reinforces the hypothesis that communicating negative information improves tracking accuracy. Teams with the lowest robot to target ratios cover the least of the environment. Making use of negative information affords the largest improvement in performance. Teams with high robot to target ratios already cover the environment well, and find little use for negative information.

7. Conclusion and Future Work

This paper has presented the theory that communication of *negative information* can improve the accuracy of multirobot, multitarget tracking. The most improvement can be seen in situations in which the observing robots are unable to cover the entire environment. It seems likely the benefits of communicating negative information is not unique to this to this particular domain. Related work such as (Jung and Sukhatme, 2002b) or (Stroupe and Balch, 2004b) might be improved in this way.

We are currently working on implementing a real-time multirobot multitarget tracking system in the form of a robot soccer team that can cooperatively track its opponents. We are also looking at ways in which this principle can be used in sensor networks to either improve position estimate accuracy or reduce the number of sensors necessary to achieve a given performance goal. It is expected that this principle will easily cross domain lines, although this has not yet been validated.

Acknowledgment

We want to acknowledge NSF grant #0347743 for funding this work.

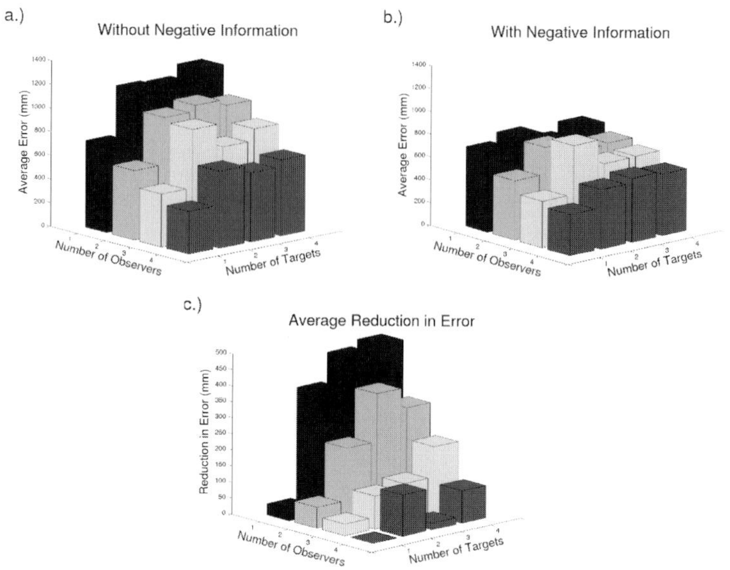

Figure 3. The results of the baseline and experimental systems. Figure 3a shows the average error of all trials of the baseline system, by number of observers and number of targets. 3b shows the average error of all trials of the experimental system. 3c shows the reduction in error by the experimental system over the baseline system. Positive values indicate a reduction in error by the experimental system.

References

Arulampalam, S., M.-S. G. N. and Clapp, T. (2002). A tutorial on particle filters for on-line non-linear/non-gaussian bayesian tracking. *IEEE Transactions on Signal Processing*.

Bar-Shalom, Y., F. T. and Scheffe, M. (1980). Joint probabilistic data association for multiple targets in clutter. In *Conference on Information Sciences and Systems*.

Bruce, J., B. T. and Veloso, M. (2000). Fast and inexpensive color image segmentation for interactive robots. In *2003 IEEE International Conference on Intelligent Robotics and Systems*.

Carpenter, J., C. P. and Fernhead, P. (1997). An improved particle filter for non-linear problems. Technical report, Department of Statistics, University of Oxford.

Dellaert, F., F. D. B. W. and Thrun, S. (1999). Monte carlo localization for mobile robots. In *IEEE International Conference on Robotics and Automation*.

Fortmann, T., B.-S. Y. and Scheffe, M. (1983). Sonar tracking of multiple targets using joint probabilistic data association. *IEEE Journal of Oceanic Engineering*, 8.

Gordon, N., S.-D. and Smith, A. (1993). Novel approach to nonlinear/non-gaussian bayesian state estimation. *Communication, Radar and Signal Processing*, (140):107–113.

Howard, A., M.-M. and Sukhatme, G. (2003). Putting the 'i' in 'team': an ego-centric approach to cooperative localization. In *IEEE International Conference on Robotics and Automation*.

Isard, M. and Blake, A. (1996). Contour tracking by stochastic propagation of conditional density. In *European Conference on Computer Vision*, pages 343–356.

Jung, B. and Sukhatme, G. (2002a). Cooperative tracking using mobile robots and environment embedded, networked sensors. In *IEEE International Symposium on Computational Intelligence in Robotics and Automation*.

Jung, B. and Sukhatme, G. (2002b). A region-based approach for coopeative multi-target tracking in a structured environment. In *IEEE International Conference Intelligent Robots and Systems*.

Jung, B. and Sukhatme, G. (2002c). Tracking targets using multiple robots: The effect of environment occlusion. *Autonoumous Robots Journal*, 13(3):191–205.

Khan, Z., B.-T. and Dellaert, F. (2003a). Efficient particle filter-based tracking of multiple interacting targets using an mrf-based motion model.

Khan, Z., B.-T. and Dellaert, F. (2003b). An mcmc-based particle filter for tracking multiple interacting targets. Technical Report GIT-GVU-03-35, College of Computing GVU Center, Georgia Institute of Technology.

Parker, L. (1997). Cooperative motion control for multi-target observation. In *IEEE International Conference Intelligent Robots and Systems*.

Parker, L. (1999). Cooperative robotics for multi-target observation. *Intelligent Automation and Soft Computing*, 5(1):5–19.

Reid, D. (1979). An algorithm for tracking multiple targets. *IEEE Trans. on Automation and Control*, AC-24:84–90.

Schulz, D., B.-W. F. D. and Cremers, A. B. (2001). Tracking multiple moving targets with a mobile robot using particle filters and statistical data association. In *IEEE International Conference on Robotics and Automation*.

Stroupe, A. (2003). *Collaborative Execution of Exploration and Tracking Using Move Value Estimation for Robot Teams*. PhD thesis, Carnegie Mellon University.

Stroupe, A. and Balch, T. (2004a). Value-based action selection for observation with robot teams using probabilistic techniques. *Journal of Robotics and Autonomous Systems*.

Stroupe, A., R.-R. and Balch, T. (2004b). Value-based action selection for observation of dynamic objects with robot teams. In *IEEE International Conference on Robotics and Automation*.

Werger, B. and Matarić, M. (2000). Broadcast of local eligibility: Behavior-based control for strongly cooperative robot teams. In *Distributed Autonomous Robotic Systems*.

ENABLING AUTONOMOUS SENSOR-SHARING FOR TIGHTLY-COUPLED COOPERATIVE TASKS

Lynne E. Parker, Maureen Chandra, Fang Tang
Distributed Intelligence Laboratory, Department of Computer Science
The University of Tennessee, Knoxville, Tennessee USA
parker@cs.utk.edu, chandra@cs.utk.edu, ftang@cs.utk.edu

Abstract This paper presents a mechanism enabling robot team members to share sensor information to achieve tightly-coupled cooperative tasks. This approach, called ASyMTRe, is based on a distributed extension of schema theory that allows schema-based building blocks to be interconnected in many ways, regardless of whether they are on the same or different robots. The inputs and outputs of schema are labeled with an information type, inspired by the theory of information invariants. By enabling robots to autonomously configure their distributed schema connections based on the flow of information through the system, robot teams with different collective capabilities are able to generate significantly different cooperative control strategies for solving the same task. We demonstrate the ability of this approach to generate different cooperative control strategies in a proof-of-principle implementation on physical robots performing a simple transportation task.

Keywords: Sensor-sharing, heterogeneous teams, multi-robot coalitions

1. Introduction

In multi-robot systems, it is advantageous to be able to treat each sensory resource on the team as a resource available to any necessitous robot team member, rather than being exclusively owned by an individual robot. The ability to share sensory information, appropriately translated to another robot's perspective, can extend the task capabilities of a given multi-robot team. In practice, however, this is difficult to achieve because each sensory resource is in fact fixed on a particular robot, and provides information only from that robot's frame of reference. Typically, mechanisms for sharing distributed sensory information are developed in an application-specific manner. The human designer might pre-define roles or subtasks, together with a list of required

L.E. Parker et al. (eds.),
Multi-Robot Systems. From Swarms to Intelligent Automata. Volume III, 119–130.

capabilities needed to accomplish each role or subtask. The robot team members can then autonomously select actions using any of a number of common approaches to multi-robot task allocation (see (Gerkey and Mataric, 2004) for a comparison of various task allocation approaches), based upon their suitability for the role or subtask, as well as the current state of the multi-robot system. The shortcoming of this approach is that the designer has to consider in advance all of the possible combinations of robot capabilities that might be present on a multi-robot team performing a given task, and to design cooperative behaviors in light of this advance knowledge.

However, as described in (Parker, 2003), the specific robot capabilities present on a team can have a significant impact on the approach a human designer would choose for the team solution. The example given in (Parker, 2003) is that of deploying a mobile sensor network, in which cooperative solutions for the same task could involve potential-field-based dispersion, marsupial delivery, or assistive navigation, depending on the capabilities of the team members.

Our research is aimed at overcoming these challenges by designing flexible sensor-sharing mechanisms within robot behavior code that do not require task-specific, pre-defined cooperative control solutions, and that translate directly into executable code on the robot team members. Some related work in sensor-sharing has led to the development of application-specific solutions that allow a robot team member to serve as a remote viewer of the actions of other teammates, providing feedback on the task status to its teammates. In particular, this has been illustrated by several researchers in the multi-robot box pushing and material handling domain (Gerkey and Mataric, 2002, Adams et al., 1995, Spletzer et al., 2001, Donald et al., 1997), in which one or more robots push an object while a remote robot or camera provides a perspective of the task status from a stand-off position. Our work is aimed at generating these types of solutions automatically, to enable robot teams to coalesce into sensor-sharing strategies that are not pre-defined in advance.

Our approach, which we call ASyMTRe (Automated Synthesis of Multi-robot Task solutions through software Reconfiguration, pronounced "Asymmetry"), is based on a combination of schema theory (Arkin et al., 1993) and inspiration from the theory of information invariants (Donald et al., 1993). The basic building blocks of our approach are collections of *perceptual schemas*, *motor schemas*, and a simple new component we introduce, called *communication schemas*. These schemas are assumed to be supplied to the robots when they are brought together to form a team, and represent baseline capabilities of robot team members. The ASyMTRe system configures a solution by choosing from different ways of combining these building blocks into a teaming solution, preferring the solution with the highest utility. Different combinations of building blocks can yield very different types of cooperative solutions to the same task.

In a companion paper (Tang and Parker, 2005), we have described an automated reasoning system for generating solutions based on the schema building blocks. In this paper, we focus on illustrating a proof-of-principle task that shows how different interconnections of these schema building blocks can yield fundamentally different solution strategies for sensor-sharing in tightly-coupled tasks. Section 2 outlines our basic approach. Section 3 defines a simple proof of principle task that illustrates the ability to formulate significantly different teaming solutions based on the schema representation. Section 4 presents the physical robot results of this proof-of-principle task. We present concluding remarks and future work in Section 5.

2. Approach

Our ASyMTRe approach to sensor-sharing in tightly-coupled cooperative tasks includes a formalism that maps environmental, perceptual, and motor control schemas to the required flow of information through the multi-robot system, as well as an automated reasoning system that derives the highest-utility solution of schema configurations across robots. This approach enables robots to reason about how to solve a task based upon the fundamental information needed to accomplish the objectives. The fundamental information will be the same regardless of the way that heterogeneous team members may obtain or generate it. Thus, robots can collaborate to define different task strategies in terms of the required flow of information in the system. Each robot can know about its own sensing, effector, and behavior capabilities and can collaborate with others to find the right combination of actions that generate the required flow of information to solve the task. The effect is that the robot team members interconnect the appropriate schemas on each robot, and across robots, to form coalitions (Shehory, 1998) to solve a given task.

2.1 Formalism of Approach

We formalize the representation of the basic building blocks in the multi-robot system as follows:

- A class of *Information*, denoted $F = \{F_1, F_2, ...\}$.

- *Environmental Sensors*, denoted $ES = \{ES_1, ES_2, ...\}$. The input to ES_i is a specific physical sensor signal. The output is denoted as $O^{ES_i} \in F$.

- *Perceptual Schemas*, denoted $PS = \{PS_1, PS_2, ...\}$. Inputs to PS_i are denoted $I_k^{PS_i} \in F$. The perceptual schema inputs can come from either the outputs of communication schemas or environmental sensors. The output is denoted $O^{PS_i} \in F$.

- *Communication Schemas*, denoted $CS = \{CS_1, CS_2, ...\}$. The inputs to CS_i are denoted $I_k^{CS_i} \in F$. The inputs originate from the outputs of perceptual schemas or communication schemas. The output is denoted $O^{CS_i} \in F$.

- *Motor Schemas*, denoted $MS = \{MS_1, MS_2, ...\}$. The inputs to MS_i are denoted $I_k^{MS_i} \in F$, and come from the outputs of perceptual schemas or communication schemas. The output is denoted $O^{MS_i} \in F$, and is connected to the robot effector control process.

- A set of n robots, denoted $R = \{R_1, R_2, ..., R_n\}$. Each robot is described by the set of schemas available to that robot: $R_i = \{ES^i, PS^i, CS^i, MS^i\}$, where ES^i is the set of environmental sensors available to R_i, and PS^i, CS^i, MS^i are the sets of perceptual, communication, and motor schemas available to R_i, respectively.

- $Task = \{MS_1, MS_2, ...\}$, which is the set of motor schemas that must be activated to accomplish the task.

A valid configuration of schemas distributed across the robot team has all of the inputs and outputs of the schemas in T connected to appropriate sources, such that the following is true: $\forall_k \exists_i CONNECT(O^{S_i}, I_k^{S_j}) \Leftrightarrow O^{S_i} = I_k^{S_j}$, where S_i and S_j are types of schemas. This notation means that for all the inputs of S_j, there exists some S_i whose output is connected to one of the required inputs. In(Tang and Parker, 2005), we define quality metrics to enable the system to compare alternative solutions and select the highest-quality solution. Once the reasoning system has generated the recommended solution, each robot activates the required schema interconnections in software.

3. Proof-of-Principle Task Implementation

To show that it is possible to define basic schema building blocks to enable distributed sensor sharing and flexible solution approaches to a tightly-coupled cooperative task, we illustrate the approach in a very simple proof of principle task. This task, which we call the *transportation task*, requires each robot on the team to navigate to its pre-assigned, unique goal point. In order for a robot to reach its assigned goal, it needs to know its current position relative to its goal position so that it can move in the proper direction. In some cases, a robot may be able to sense its current position using its own sensors. In other cases, the robot may not have enough information to determine its current position. In the latter case, other more capable robots can help by sharing sensor information with the less capable robot.

As shown in Table 1, the environmental sensors available in this case study are a laser scanner, a camera, and Differential GPS. A robot can use a laser

Table 1. Environmental Sensors (ES) and Robot Types for proof-of-principle task.

Environmental Sensors			Robot Types	
Name	Description	Info. Type	Name	Available Sensors
ES_1	Laser	*laserscanner*	R_1	Laser
ES_2	Camera	*ccd*	R_2	Camera
ES_3	DGPS	*dgps*	R_3	DGPS
			R_4	Laser and Camera
			R_5	Laser and DGPS
			R_6	Camera and DGPS
			R_7	Laser and Camera and DGPS
			R_8	—

Table 2. Perceptual and Communications Schemas for proof-of-principle task.

Perceptual Schemas		
Name	**Input Info. Type**	**Output Info. Type**
PS_1	*laserrange* OR *dgps* OR *curr-global-pos(self)* OR (*curr-rel-pos(other$_k$)* AND *curr-global-pos(other$_k$)*)	*curr-global-pos(self)*
PS_2	—	*curr-global-goal(self)*
PS_3	(*curr-global-pos(self)* AND *curr-rel-pos(other$_k$)*)	*curr-global-pos(other$_k$)*
PS_4	*laserrange* or *ccd*	*curr-rel-pos(other$_k$)*
PS_5	*curr-global-pos(other)*	*curr-global-pos(other)*

Communication Schemas		
Name	**Input Info. Type**	**Output Info. Type**
CS_1	*curr-global-pos(self)*	*curr-global-pos(other$_k$)*
CS_2	*curr-global-pos(other$_k$)*	*curr-global-pos(self)*

scanner with an environmental map to localize itself and calculate its current global position. A robot's camera can be used to detect the position of another robot relative to itself. The DGPS sensor can be used outdoors for localization and to detect the robot's current global position. Based upon these environmental sensors, there are eight possible combinations of robots, as shown in Table 1. In this paper, we focus on three types of robots – R_8: a robot that possesses no sensors; R_2: robot that possesses only a camera; and R_4: a robot that possesses a camera and a laser ranger scanner (but no DGPS).

For this task, we define five perceptual schemas, as shown in Table 2. PS_1 calculates a robot's current global position. With the sensors we have defined, this position could be determined either by using input data from a laser scanner combined with an environmental map, from DGPS, or from communication schemas supplying similar data. For an example of this latter case, a robot can calculate its current global position by knowing the global position of another robot, combined with its own position relative to the globally positioned robot. PS_2 outputs a robot's goal position, based on the task definition provided by the user. PS_3 calculates the current global position of a remote robot based on two inputs – the position of the remote robot relative to itself and its own current global position. PS_4 calculates the position of another robot relative to itself. Based on the defined sensors, this calculation could be derived from either a laser scanner or a camera. PS_5 receives input from another robot's communication schema, CS_1, which communicates the current position of that other robot.

Communication schemas communicate data to another robot's perceptual schemas. As shown in Table 2, CS_1 communicates a robot's current global position to another robot, while CS_2 communicates the current global position of a remote robot that remote robot. Motor schemas send control signals to the robot's effectors to enable the robot to accomplish the assigned task. In this case study, we define only one motor schema, MS, which encodes a *go-to-goal* behavior.

The input information requirements of MS are *curr-global-pos*(*self*) and *curr-global-goal*(*self*). In this case, the motor schema's output is derived based on the robot's current position received from PS_1 and its goal position received from PS_2.

Figure 1 shows all the available schemas for this task and how they can be connected to each other, based on the information labeling. The solid-line arrows going into a schema represent an "OR" condition – it is sufficient for the schema to only have one of the specified inputs to produce output. The dashed-line arrows represent an "AND" condition, meaning that the schema requires all of the indicated inputs for it to calculate an output. For example, PS_1 can produce output with input(s) from either ES_1 (combined with the environmental Map), ES_3, CS_2^j (R_j's CS_2), or (PS_4 and PS_5).

4. Physical Robot Experiments

These schema were implemented on two Pioneer robots equipped with a SICK laser range scanner and a Sony pan-tilt-zoom camera. Both robots also possessed a wireless ad hoc networking capability, enabling them to communicate with each other. Experiments were conducted in a known indoor environment using a map generated using an autonomous laser range mapping algorithm. Laser-based localization used a standard Monte-Carlo Localization technique. The code for the implementation of PS_4 makes use of prior work by (Parker et al., 2004) for performing vision-based sensing of the relative position of another robot. This approach makes use of a cylindrical marker designed to provide a unique robot ID, as well as relative position and orienta-

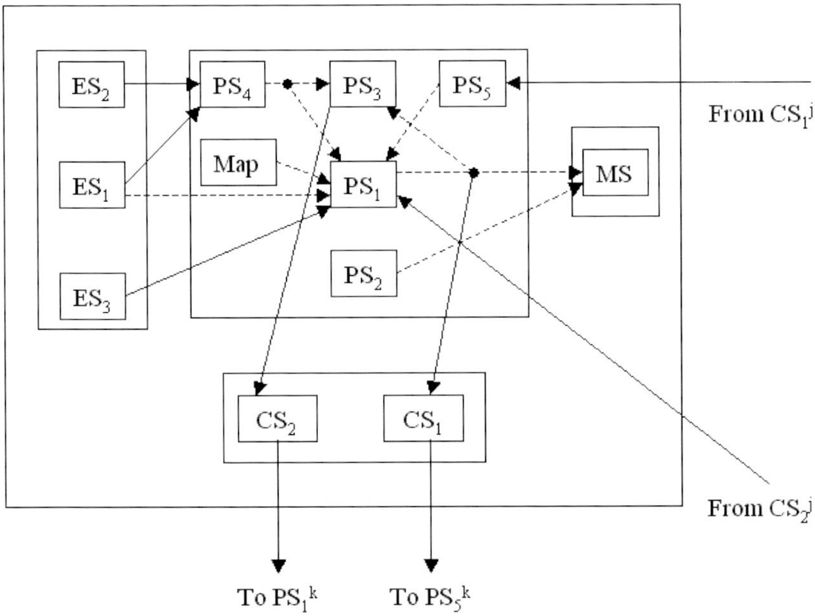

Figure 1. Illustration of connections between all available schemas.

tion information suitable for a vision-based analysis. Using these two robots, three variations on sensor availability were tested to illustrate the ability of these building blocks to generate fundamentally different cooperative behaviors of the same task through sensor sharing. In these experiments, the desired interconnections of schemas were developed by hand; in subsequent work, we can now generate the required interconnections automatically through our ASyMTRe reasoning process (Tang and Parker, 2005).

Variation 1. The first variation is a baseline case in which both robots are of type R_4, meaning that they have full use of both their laser scanner and a camera. Each robot localizes itself using its laser scanner and map and reaches its own unique goals independently. This case is the most ideal solution but only works if the both robots possess laser scanners and maps. If one of the robots loses its laser scanner, this solution no longer works. Figure 2 shows the schema instantiated on the robots for this variation. PS_1 and PS_2 are connected to MS to supply the required inputs to the *go-to-goal* behavior. Also shown in Figure 2 are snapshots of the robots performing this instantiation of the schema. In this case, since both robots are fully capable, they move towards their goals independently without the need for any sensor sharing or communication.

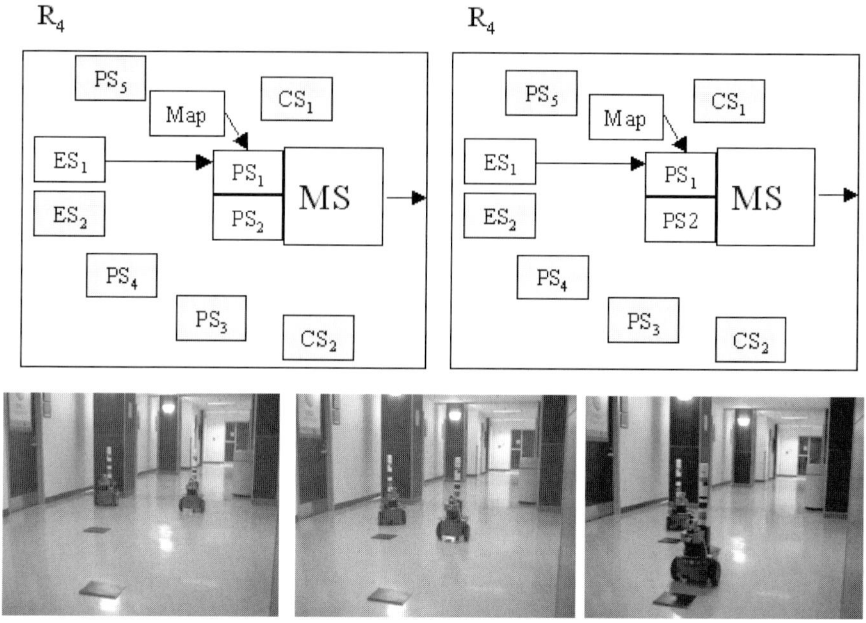

Figure 2. Results of Variation 1: Two robots of type R_4 performing the task without need for sensor-sharing or communication. Goals are black squares on the floor. Graphic shows schema interconnections (only white boxes activated).

Variation 2. The second variation involves a fully capable robot of type R_4, as well as a robot of type R_2 whose laser scanner is not available, but still has use of its camera. As illustrated in Figure 3, Robot R_4 helps R_2 by communicating (via CS_1) its own current position, calculated by PS_1 using its laser scanner (ES_1) and environmental map. Robot R_2 receives this communication via PS_5 and then uses its camera (ES_2) to detect R_4's position relative to itself (via PS_4) and calculate its own current global position (using PS_1) based on R_4's relative position and R_4's communicated global position. Once both robots know their own current positions and goal positions, their motor schemas can calculate the motor control required to navigation to their goal points. Figure 3 also shows snapshots of the robots performing the Variation 2 instantiation of the schema. In this case, R_2 begins by searching for R_4 using its camera. At present, we have not yet implemented the constraints for automatically ensuring that the correct line of sight is maintained, so we use communication to synchronize the robots. Thus, when R_2 locates R_4, it communicates this fact to R_4. R_4 then is free to move towards its goal. If R_2 were to lose sight of R_4, it would communicate a message to R_4 to re-synchronize the relative sighting of R_4 by

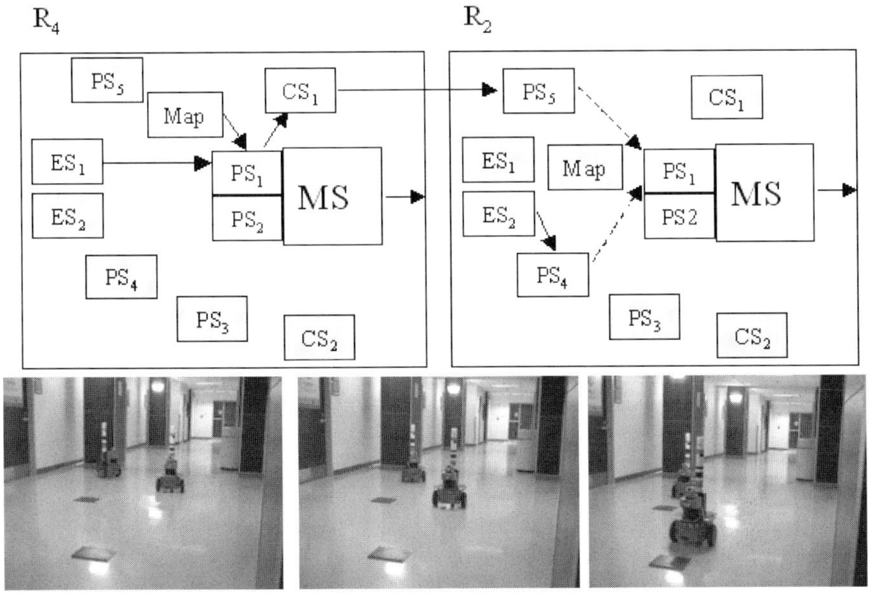

Figure 3. Variation 2: A robot of type R_4 and of type R_2 share sensory information to accomplish their task. Here, R_2 (on the left) turns toward R_4 to localize R_4 relative to itself. R_4 communicates its current global position to R_2, enabling it to determine its own global position, and thus move successfully to its goal position.

R_2. With this solution, the robots automatically achieve navigation assistance of a less capable robot by a more capable robot.

Variation 3. The third variation involves a sensorless robot of type R_8, which has access to neither its laser scanner nor camera. As illustrated in Figure 4, the fully-capable robot of type R_4 helps R_8 by communicating R_8's current global position. R_4 calculates R_8's current global position by first using its own laser (ES_1) and map to calculate its own current global position (PS_1). R_4 also uses its own camera (ES_2) to detect R_8's position relative to itself (using PS_4). Then, based on this relative position and its own current global position, R_4 calculates R_8's current global position (using PS_3) and communicates this to R_8 (via CS_2). Robot R_8 feeds its own global position information from R_4 directly to its motor schema. Since both of the robots know their own current and goal positions, each robot can calculate its motor controls for going to their goal positions. Figure 4 also shows snapshots of the robots performing the Variation 3 instantiation of the schema. With this solution, the robots automatically achieve navigation assistance of a sensorless robot by a more capable robot.

Analysis. In extensive experimentation, data on the time for task completion, communication cost, sensing cost, and success rate was collected as an average

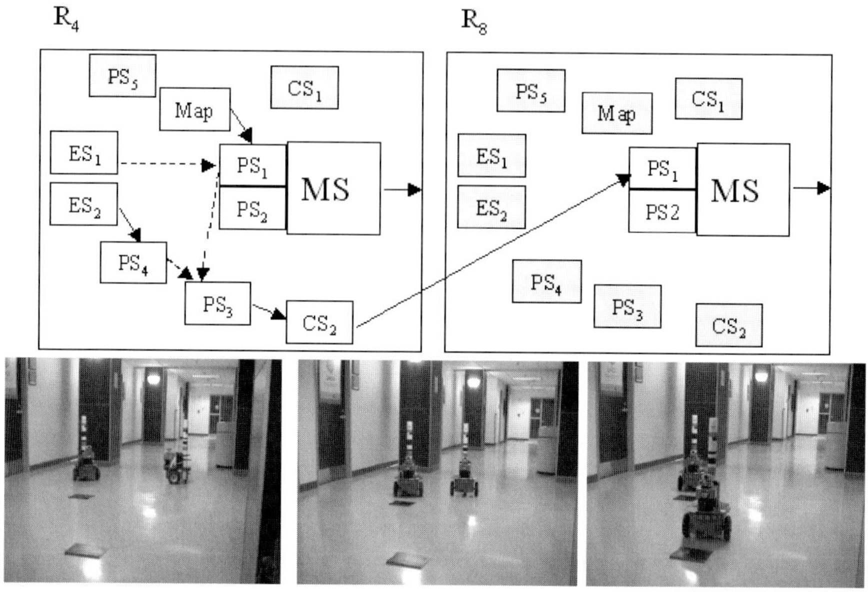

Figure 4. Variation 3: A robot of type R_4 helps a robot with no sensors (type R_8) by sharing sensory information so that both robots accomplish the objective. Note how R_4 (on the right) turns toward R_8 to obtain vision-based relative localization of R_8. R_4 then guides R_8 to its goal position. Once R_8 is at its goal location, R_4 then moves to its own goal position.

of 10 trials of each variation. Full details are available in (Chandra, 2004). We briefly describe here the success rate of each variation. In all three variations, robot R_4 was 100% successful in reaching its goal position. Thus, for Variation 1, since the robots are fully capable and do not rely on each other, the robots always succeeded in reaching their goal positions. In Variation 2, robot R_2 succeeded in reaching its goal 6 times out of 10, and in Variation 3, robot R_8 successfully reached its goal 9 times out of 10 tries. The failures of robots R_4 and R_8 in Variations 2 and 3 were caused by variable lighting conditions that led to a false calculation of the relative robot positions using the vision-based robot marker detection. However, even with these failures, these overall results are better than what would be possible without sensor sharing. In Variations 2 and 3, if the robots did not share their sensory resources, one of the robots would never reach its goal position, since it would not have enough information to determine its current position. Thus, our sensor sharing mechanism extends the ability of the robot team to accomplish tasks that otherwise could not have been achieved.

5. Conclusions and Future Work

In this paper, we have shown the feasibility of the ASyMTRe mechanism to achieve autonomous sensor-sharing of robot team members performing a tightly-coupled task. This approach is based on an extension to schema theory, which allows schemas distributed across multiple robots to be autonomously connected and executed at run-time to enable distributed sensor sharing. The inputs and outputs to schemas are labeled with unique information types, inspired by the theory of information invariants, enabling any schema connections with matching information types to be configured, regardless of the location of those schema or the manner in which the schema accomplishes its job. We have demonstrated, through a simple transportation task implemented on two physical robots, the ability of the schema-based methodology to generate very different cooperative control techniques for the same task based upon the available sensory capabilities of the robot team members. If robots do not have the ability to accomplish their objective, other team members can share their sensory information, translated appropriately to another robot's frame of reference. This approach provides a framework within which robots can generate the highest-quality team solution for a tightly-coupled task, and eliminates the need of the human designer to pre-design all alternative solution strategies. In continuing work, we are extending the formalism to impose motion constraints (such as line-of-sight) needed to ensure that robots can successfully share sensory data while they are in motion, generalizing the information labeling technique, and implementing this approach in more complex applications. In addition, we are developing a distributed reasoning approach that enables team members to autonomously generate the highest-quality configuration of schemas for solving the given task.

References

Adams, J. A., Bajcsy, R., Kosecka, J., Kumar, V., Mandelbaum, R., Mintz, M., Paul, R., Wang, C., Yamamoto, Y., and Yun, X. (1995). Cooperative material handling by human and robotic agents: Module development and system synthesis. In *Proc. of IEEE/RSJ International Conference on Intelligent Robots and Systems*.

Arkin, R. C., Balch, T., and Nitz, E. (1993). Communication of behavioral state in multi-agent retrieval tasks. In *Proceedings of the 1993 International Conference on Robotics and Automation*, pages 588–594.

Chandra, M. (2004). Software reconfigurability for heterogeneous robot cooperation. Master's thesis, Department of Computer Science, University of Tennessee.

Donald, B. R., Jennings, J., and Rus, D. (1993). Towards a theory of information invariants for cooperating autonomous mobile robots. In *Proceedings of the International Symposium of Robotics Research*, Hidden Valley, PA.

Donald, B. R., Jennings, J., and Rus, D. (1997). Information invariants for distributed manipulation. *International Journal of Robotics Research*, 16(5):673–702.

Gerkey, B. and Mataric, M. (2002). Pusher-watcher: An approach to fault-tolerant tightly-coupled robot cooperation. In *Proc. of IEEE International Conference on Robotics and Automation*, pages 464–469.

Gerkey, B. and Mataric, M. J. (2004). A formal analysis and taxonomy of task allocation in multi-robot systems. *Int. J. of Robotics Research*, 23(9):939–954.

Parker, L. E. (2003). The effect of heterogeneity in teams of 100+ mobile robots. In Schultz, A., Parker, L. E., and Schneider, F., editors, *Multi-Robot Systems Volume II: From Swarms to Intelligent Automata*. Kluwer.

Parker, L. E., Kannan, B., Tang, F., and Bailey, M. (2004). Tightly-coupled navigation assistance in heterogeneous multi-robot teams. In *Proceedings of IEEE International Conference on Intelligent Robots and Systems*.

Shehory, O. (1998). Methods for task allocation via agent coalition formation. *Artificial Intelligence*, 101(1-2):165–200.

Spletzer, J., Das, A., Fierro, R., Tayler, C., Kumar, V., and Ostrowski, J. (2001). Cooperative localization and control for multi-robot manipulation. In *Proc. of the IEEE/RSJ International Conference on Intelligent Robots and Systems*, Hawaii.

Tang, F. and Parker, L. E. (2005). ASyMTRe: Automated synthesis of multi-robot task solutions through software reconfiguration. *Submitted*.

IV

DISTRIBUTED MAPPING AND COVERAGE

MERGING PARTIAL MAPS WITHOUT USING ODOMETRY

Francesco Amigoni*, Simone Gasparini
Dipartimento di Elettronica e Informazione
Politecnico di Milano, 20133 Milano, Italy
amigoni@elet.polimi.it, gasparin@elet.polimi.it

Maria Gini
Dept of Computer Science and Engineering
University of Minnesota, Minneapolis, USA†
gini@cs.umn.edu

Abstract Most map building methods employed by mobile robots are based on the assumption that an estimate of the position of the robots can be obtained from odometry readings. In this paper we propose methods to build a global geometrical map by integrating partial maps without using any odometry information. This approach increases the flexibility in data collection. Robots do not need to see each other during mapping, and data can be collected by a single robot or multiple robots in one or multiple sessions. Experimental results show the effectiveness of our approach in different types of indoor environments.

Keywords: scan matching, map merging, distributed mapping.

1. Introduction

In this paper we show how to build a global map of an environment by merging partial maps without using any relative position information but relying only on geometrical information. The maps we consider are collections of segments obtained from 2D laser range data. They provide a compact and easy-to-use (for example, to plan a path) representation of the environment. No hypothesis is made about the environment to be mapped: experiments demonstrate that our methods work well both in regular and in scattered environments. We reduce the merging of a sequence of partial maps to the it-

*Partial funding provided by a Fulbright fellowship and by "Progetto Giovani Ricercatori" 1999.
†Partial funding provided by NSF under grants EIA-0224363 and EIA-0324864.

L.E. Parker et al. (eds.),
Multi-Robot Systems. From Swarms to Intelligent Automata. Volume III, 133–144.

erated integration of two partial maps. The methods we propose are robust to displacements between the partial maps, provided that they have at least a common geometrical feature.

Map merging without odometry information has the advantage of being independent from how the data have been collected. It is indifferent if the partial maps are collected during a single session or multiple sessions, by a single robot or multiple robots. Robots can be added or removed at any time, and they do not need to know their own position. For the experiments in this paper we used a single robot but all the results are applicable to multirobots.

This paper is structured as follows. The next section discusses the main approaches to scan matching and map merging. Section 3 describes our scan matching algorithm, and Section 4 our map merging methods. Experimental results are in Section 5.

2. Previous Work

Scan matching is the process of calculating the translation and rotation of a scan to maximize its overlap with a reference scan. A number of scan matching algorithms have been presented in the last two decades; they differ for the kind of environments in which they perform well, e.g., with straight perpendicular walls (Weiss et al., 1994), for the computational effort, for the choice of operating directly on the scan data (Lu and Milios, 1997) or on segments approximating the data (Zhang and Ghosh, 2000). All these methods require an initial position estimate to avoid erroneous alignments of the two scans.

The most used algorithm (Lu and Milios, 1997) iteratively minimizes an error measure by first finding a correspondence between points in the two scans, and then doing a least square minimization of all point-to-point distances to determine the best transformation. When the environment consists of straight perpendicular walls matching is simpler. Cross-correlation of the histograms of angles between the actual and previous scans provides the orientation of the two maps, while the translation is obtained either by cross-correlation of the distance histogram (Weiss et al., 1994) or by least square minimization (Martignoni III and Smart, 2002). These methods are sensitive to large displacements between the maps and to changes in the environment.

Map merging, namely the problem of building a global map from data collected by several robots, is usually solved by extending SLAM techniques (Burgard et al., 2002, Ko et al., 2003, Fenwick et al., 2002), or using EM (Simmons et al., 2000, Thrun et al., 2000).

Most map merging techniques rely on the assumption that the robot positions are known. For example, in (Simmons et al., 2000, Burgard et al., 2002) the positions of the robots are assumed to be known at all times; in (Thrun et al., 2000) the robots don't know their relative start position but each robot has to start within sight of the team leader. An exception is the work in (Konolige et al., 2003) where map merging is done using a decision theoretic approach. The robots do not need to know their own position, but to facilitate the match the maps have to be annotated manually with distinctive features. In (Ko et al., 2003), particle filters are used for partial map localization and the robots have to actively verify their relative locations before the maps are merged.

3. Method for Scan Matching

We propose a MATCH function for matching two partial maps composed of segments. Our method is exclusively based on the geometrical information and constraints (Grimson, 1990) contained in the partial maps. In particular, we consider angles between pairs of segments in the partial maps as a sort of "geometrical landmarks" on which the matching process is based. This use of "local" geometrical features is significantly different from other related works in map building that use "global" geometrical features (e.g., those represented by an histogram of angle differences). MATCH integrates two partial maps into a third one. Let's call S_1 and S_2 the two partial maps and $S_{1,2}$ the resulting map. MATCH operates in three major steps:

1. Determine possible transformations. This step first finds the angles between the segments in S_1 and between the segments in S_2 and then finds the possible transformations (namely, the rotations and translations) that superimpose at least one angle α_2 of S_2 to an (approximately) equal angle α_1 of S_1. Recall that angles between pairs of segments in a partial map are the geometrical landmarks we adopt.

In principle, without any information about the relative positions of the two scans, there are $O(n_1^2 n_2^2)$ possible transformations, where n_1 and n_2 are the number of segments in S_1 and S_2, respectively. We have devised three heuristics to speed up the computation:

a *Consider Angles between Consecutive Segments.* In each scan, we consider only the angles between two consecutive segments; let A_1^s and A_2^s be the sets of such angles for S_1 and S_2, respectively. The number of possible transformations drops to $O(n_1 n_2)$. Finding the sets A_1^s and A_2^s is easy when the segments in S_1 and in S_2 are ordered, which is usually the case with laser range scanners.

b *Consider Angles between Randomly Selected Segments.* In each scan, we examine the angles between pairs of segments selected randomly. We assign a higher probability to be selected to longer segments, since they provide more precise information about the environment. Let A_1^r and A_2^r be the sets of the selected angles for S_1 and S_2, respectively. The number of transformations generated by this method is $O(a_1 a_2)$, where

$a_1 = |A_1^r|$ and $a_2 = |A_2^r|$ are the number of selected angles in A_1^r and A_2^r, respectively.

c *Consider Angles between Perpendicular Segments.* In each scan, we select only angles between perpendicular segments. This heuristic is particularly convenient for indoor environments, where walls are often normal to each other. The heuristic is computed from the histogram of segments grouped by orientation. The direction where the sum of the lengths of the segments is maximal is the *principal direction*. In Fig. 1, the histogram of a scan taken in an indoor environment is shown. The principal direction is the element L_9 and the normal direction is the element L_0. Let A_1^h and A_2^h be the sets of angles formed by a segment in the principal direction and a segment in the normal direction of the histograms of S_1 and S_2, respectively. The set of possible transformations is found comparing the angles in A_1^h and A_2^h. The number of possible transformations is $O(p_1 n_1 p_2 n_2)$, where p_i and n_i are respectively the number of segments in the principal and in the normal directions of the histogram of scan S_i.

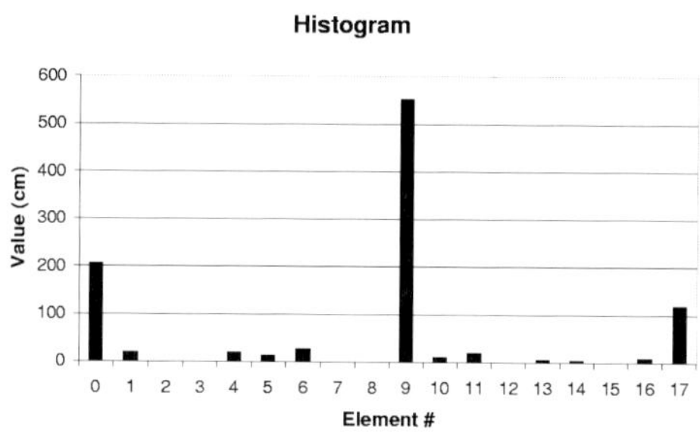

Figure 1. The histogram of a scan

2. Evaluate the transformations. To measure the goodness of a transformation t we transform S_2 on S_1 (in the reference frame of S_1) according to t (obtaining S_2^t), then we calculate the approximate length of the segments of S_1 that correspond to (namely, match with) segments of S_2^t. Thus, the measure of a transformation is the sum of the lengths of the corresponding segments that the transformation produces. More precisely, for every pair of segments $s_1 \in S_1$ and $s_2^t \in S_2^t$ we project s_2^t on the line supporting s_1 and compute the length l_1 of the common part of $s1$ and the projected segment. We repeat the process by projecting s_1 on s_2^t, obtaining l_2. The average of l_1 and l_2 is a measure of how the pair of segments match. This step evaluates a single transformation by considering all the pairs of segments of the two scans that are $O(n_1 n_2)$.

3. Apply the best transformation and fuse the segments. Once the best transformation $\bar{t}$ has been found, this step transforms the second partial map S_2 in the reference frame of S_1 according to $\bar{t}$ obtaining $S_2^{\bar{t}}$. The map that constitutes the output of MATCH is then obtained by fusing the segments of S_1 with the segments of $S_2^{\bar{t}}$. The main idea behind the fusion of segments is that a set of matching segments is substituted in the final map by a single polyline. We iteratively build a sequence of approximating polylines $P_0, P_1, \ldots$ that converges to the polyline P that adequately approximates (and substitutes in the resulting map) a set of matching segments. The polyline P_0 is composed of a single segment connecting the pair of farthest points of the matching segments. Given the polyline P_{n-1}, call s the (matching) segment at maximum distance from its corresponding (closest) segment $\bar{s}$ in P_{n-1}. If the distance between s and $\bar{s}$ is less than the acceptable error, then P_{n-1} is the final approximation P. Otherwise, s substitutes $\bar{s}$ in P_{n-1} and s is connected to the two closest segments in P_{n-1} to obtain the new polyline P_n.

4. Methods for Map Merging

We now describe methods for integrating a sequence $S_1, S_2, \ldots S_n$ of n partial maps by repeatedly calling MATCH.

Sequential Method. This is the simplest method, which operates as follows. The first two partial maps are integrated, the obtained map then is grown by sequentially integrating the third partial map, the fourth partial map, and so on. Eventually, the final map $S_{1,2,\ldots,n}$ is constructed. In order to integrate n partial maps, the sequential method requires $n-1$ calls to MATCH. A problem with this method is that, as the process goes on, MATCH is applied to a partial map that grows larger and larger (it contains more and more segments). This will cause difficulties in the integration of S_i with large i, since S_i could match with different parts of the larger map.

Tree Method. To overcome the above problem, the integration of a small partial map with a large partial map should be avoided. This is the idea underlying the *tree* method, which works as follows. Each partial map of the initial sequence is integrated with the successive partial map of the sequence to obtain a new sequence $S_{1,2}$, $S_{2,3}$, $\ldots$, $S_{n-1,n}$ of $n-1$ partial maps. Then, each partial map of this new sequence is integrated with the successive one to obtain a new sequence $S_{1,2,3}$, $S_{2,3,4}$, $\ldots$, $S_{n-2,n-1,n}$ of $n-2$ partial maps. The process continues until a single final map $S_{1,2,\ldots,n}$ is produced. The tree method always integrates partial maps of the same size, since they approximately contain the same number of segments. The number of calls to MATCH required by the tree method to integrate a sequence of n partial maps is $n(n-1)/2$. Note also that, while the sequential method can be applied in an on-line fashion (i.e., while the robot is moving), the most natural implementation of the tree method is off-line. To speed up the tree method we have developed an heuristic that, given a sequence of partial maps at any level of the tree (let us suppose at level 0 for simplicity), attempts to integrate the partial maps S_i and S_{i+2}; if the integration succeeds, the final result $S_{i,i+2}$ represents the same map that would have been obtained with three integrations: S_i with S_{i+1} to obtain $S_{i,i+1}$, S_{i+1} with S_{i+2} to obtain $S_{i+1,i+2}$, and $S_{i,i+1}$ with $S_{i+1,i+2}$ to obtain $S_{i,i+1,i+2}$. Moreover, the number of partial maps in the new sequence is reduced by one unit, because $S_{i,i+2}$ substitutes both $S_{i,i+1}$ and $S_{i+1,i+2}$. This heuristic is best used when the partial maps S_i and S_{i+2} are already the result of a number of integrations performed by the tree method and their common part is significant. For example, in the sequence produced at the level 3 of the tree technique the first ($S_{1,2,3,4}$) and the third ($S_{3,4,5,6}$) partial maps have a significant common part, since approximately half of the two partial maps overlaps.

A problem with the tree method is due to the presence of "spurious" segments in the integrated maps, namely segments that correspond to the same part of the real environment but that are not fused together. This problem is exacerbated in the tree method since the same parts of the partial maps are repeatedly fused together.

Pivot Method. The *pivot* method combines the best features of the two previous methods. This method starts as the tree method and constructs a sequence $S_{1,2}$, $S_{2,3}$, …, $S_{n-1,n}$ of $n-1$ partial maps. At this point, we note that S_2 is part of both $S_{1,2}$ and $S_{2,3}$ and that the transformation $\bar{t}_{1,2}$ used to integrate S_1 and S_2 provides the position and orientation of the reference frame of S_2 in the reference frame of $S_{1,2}$. It is therefore possible to transform $S_{2,3}$ according to $\bar{t}_{1,2}$ and fuse the segments of the partial maps $S_{1,2}$ and $S_{2,3}^{\bar{t}_{1,2}}$ to obtain $S_{1,2,3}$. In a similar way, $S_{1,2,3,4}$ can be obtained from $S_{1,2,3}$ and $S_{3,4}$ by applying to the latter the transformation $\bar{t}_{2,3}$ and fusing the segments of $S_{1,2,3}$ and $S_{3,4}^{\bar{t}_{2,3}}$. Iterating this process, from the sequence $S_{1,2}$, $S_{2,3}$, …, $S_{n-1,n}$ the final map $S_{1,2,\ldots,n}$ is obtained. The pivot method integrates partial maps of the same size, like the tree method, and requires $n-1$ calls to MATCH, like the sequential method. (In addition it requires $n-2$ executions of the not-so-expensive step 3 of MATCH.) The pivot method is naturally implementable in an on-line system. The problem of spurious segments is reduced but not completely eliminated; a way to further reduce this problem is to fuse not $S_{1,2}$ and $S_{2,3}^{\bar{t}_{1,2}}$, but $S_{1,2}$ and $S_3^{\bar{t}_{1,3}}$, where $\bar{t}_{1,3}$ is the composition of $\bar{t}_{1,2}$ and $\bar{t}_{2,3}$.

5. Experimental Results

The described methods have been validated using a Robuter mobile platform equipped with a SICK LMS 200 laser range scanner mounted in the front of the robot at a height of approximately 50 cm. For these experiments we acquired 32 scans with angular resolution of 1° and with angular range of 180°. Each scan has been processed to approximate the points returned by the sensor with segments, according to the algorithm in (Gonzáles-Baños and Latombe, 2002). The programs have been coded in ANSI C++ employing the LEDA libraries 4.2 and they have been run on a 1 GHz Pentium III processor with Linux SuSe 8.0.

The scans have been acquired in different environments (forming a loop about 40 m long) by driving the robot manually and without recording any odometric information. We started from a laboratory, a very scattered environment, then we crossed a narrow hallway with rectilinear walls to enter a department hall, a large open space with long perpendicular walls, and finally we closed the loop re-entering the laboratory (see the dashed path in Fig. 6). The correctness of integrations has been determined by visually evaluating the starting partial maps and the final map with respect to the real environment.

5.1 Scan Matching Experiments

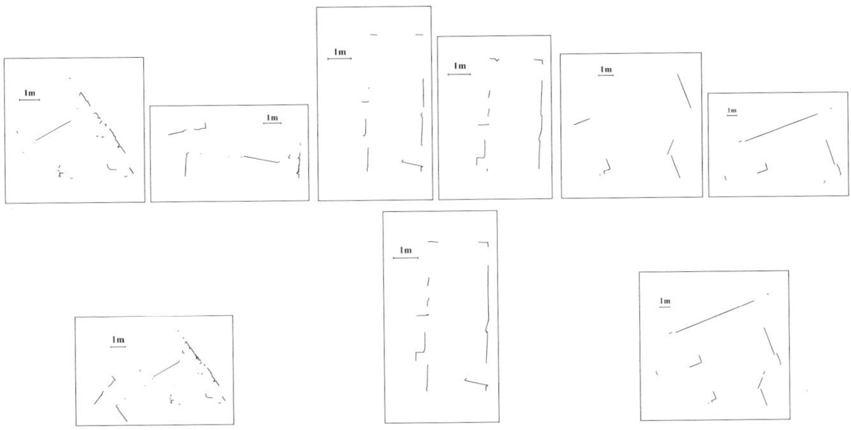

Figure 2. Top, left to right: scans S_4, S_5, S_{18}, S_{19}, S_{25}, and S_{26}; bottom, left to right: final maps $S_{4,5}$, $S_{18,19}$, and $S_{25,26}$

Table 1 shows the results obtained by matching three interesting pairs of scans (see also Fig. 2). S_4 and S_5 were taken inside the laboratory: they contain a large number of short segments since the environment is highly scattered. S_{18} and S_{19} were taken along the hallway: they contain fewer segments than the previous scans and are characterized by long rectilinear segments. S_{25} and S_{26} were taken in the hall: they contain only few segments since the environment is characterized by long rectilinear and perpendicular walls.

In general, our experimental results demonstrate that scan matching performs well (Table 2), but not all the pairs can be matched. 28 pairs of scans

Table 1. Experimental results on matching pairs of scans

Scans	S_4	S_5	S_{18}	S_{19}	S_{25}	S_{26}
# of segments	47	36	24	24	10	12
	Time	# tried	Time	# tried	Time	# tried
All transformations	936	41,260	32	3,096	0.38	231
Consecutive segments	1.25	2	0.73	27	0.13	4
Random segments	7.69	20,000	2.51	20,000	0.78	20,000
Histogram	3.29	73	1.97	192	0.15	32

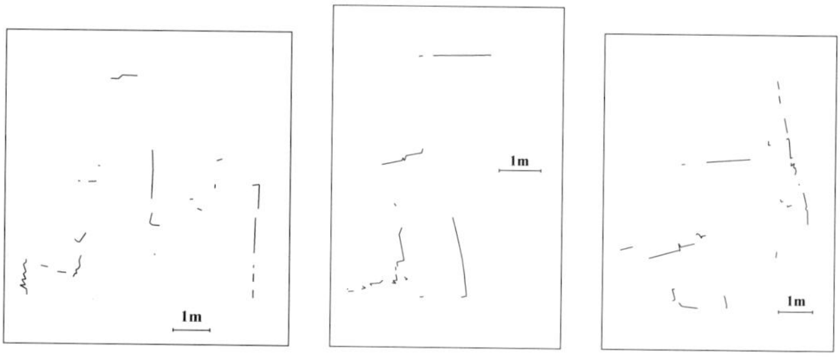

Figure 3. Scans S_1 (left), S_2 (center), and S_3 (right) taken in the lab entrance

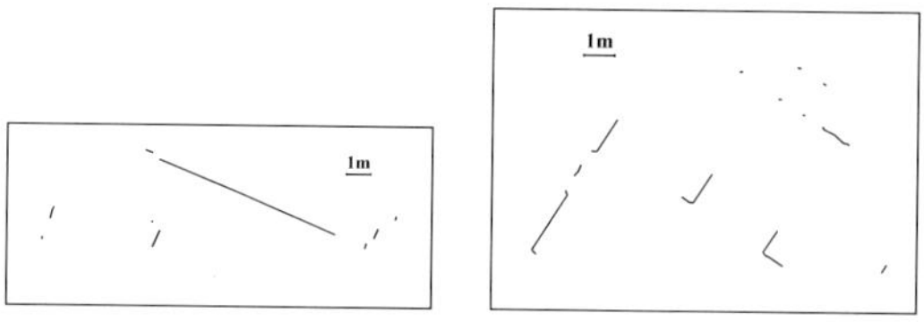

Figure 4. Scans S_{27} (left) and S_{28} (right) taken in the hall

out of 31 have been correctly matched. Unsurprisingly, the histogram-based heuristic worked well with scans containing long and perpendicular segments, as those taken in the hallway and in the hall. The heuristic based on consecutive segments seems to work well in all three kinds of environment, even if sometimes it needs some parameter adjustments.

For scan pairs $S_1 - S_2$ and $S_2 - S_3$ our method was not able to find the correct transformation. As shown in Fig. 3, the scans are extremely rich of short segments representing scattered small objects (chairs, tables, robots, and boxes). It is almost impossible, even for a human being, to find the correct match between these scans without any prior information about their relative positions. Similar problems emerged in the hall. Fig. 4 shows scans S_{27} and S_{28}, where the second one has been taken after rotating the robot of about 100 degrees. Since the environment is large and has only few objects that can be used as ref-

Table 2. Results of scan matching trials using different heuristics

	Successes	Failures
All transformations	13 (41.9%)	18 (58.1%)
Consecutive segments	21 (67.7%)	10 (32.3%)
Random segments	10 (32.2%)	21 (67.8%)
Histogram	9 (29%)	22 (71%)

erence, a drastic change of the field of view eliminates any common reference between scans, thus automatic matching is impossible.

5.2 Map Merging Experiments

We considered the sequence composed of 29 scans $S_1, S_2, \ldots, S_{29}$ (Table 3). The integration of this sequence of partial maps has been done off-line to test and compare the three methods. In all the three methods, problems arose when integrating the sub-sequence from S_{25} to S_{27} which represents the hall (Fig. 4). Here, due to a drastic rotation (about 100 degrees) of the robot in such an open and large environment, the partial maps have only one or two segments in common. In order to close the loop and complete the experiments these partial maps were manually integrated together in all the three methods.

Table 3. Experimental sequence of partial maps (the segment lengths are in mm)

Environment	Partial maps	Avg # of segments	Avg length of segments
Laboratory	$S_1 - S_7, S_{28} - S_{29}$	38.1	259.3
Hallway	$S_8 - S_{22}$	19.3	366.3
Hall	$S_{23} - S_{27}$	15.6	607
Total	$S_1 - S_{29}$	24.53	374.5

Fig. 5 shows the final map (composed of 278 segments) obtained with the sequential method. The sequential method could not integrate all the partial maps in order to close the loop: the method suddenly failed when we tried to integrate S_{21}, which has only a few short segments in common with the rest of the map.

Fig. 6 shows the final map (composed of 519 segments) obtained with the tree method. We applied the standard tree method until level 3 of the tree, then we applied the heuristic presented in Section 4 to speed up the process. As we went down in the tree, the size of the maps grew larger and larger and the execution of MATCH slowed down. For example, the integration of two partial maps (composed of 108 and 103 segments) at level 3 of the tree requires

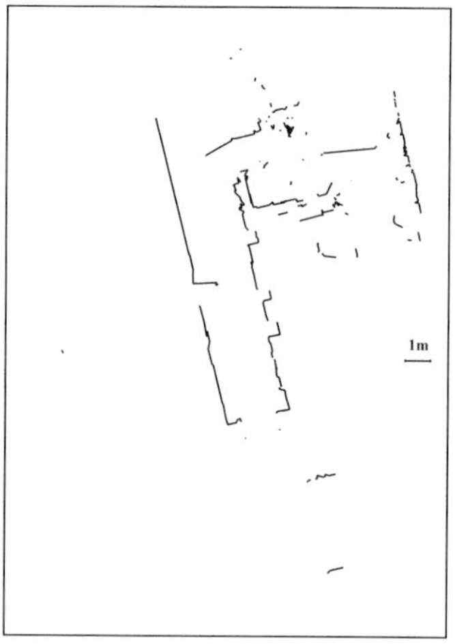

Figure 5. The final map obtained with the sequential method. ©2004 by IEEE (Amigoni et al., 2004)

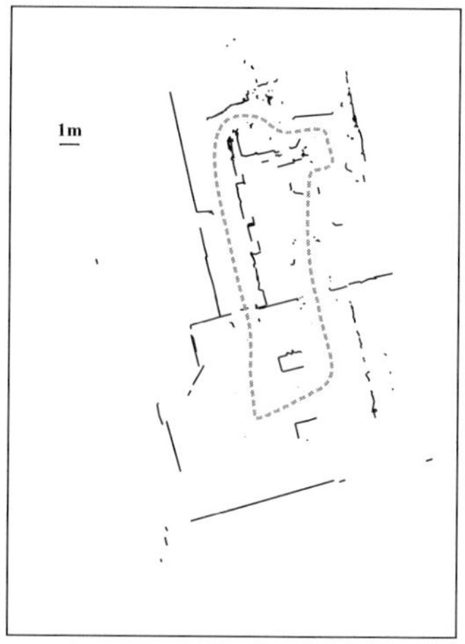

Figure 6. The final map obtained with the tree method. ©2004 by IEEE (Amigoni et al., 2004)

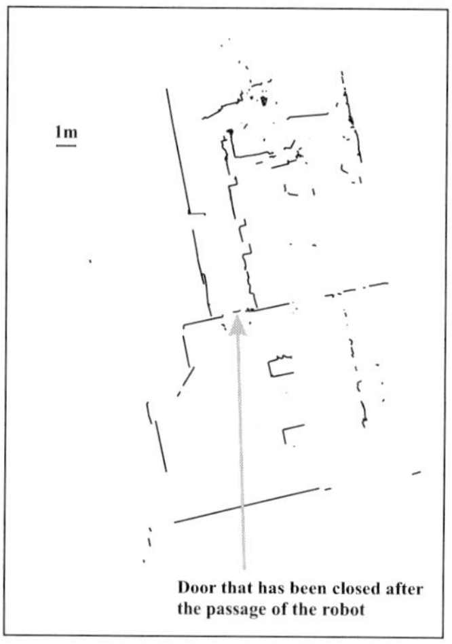

Figure 7. The final map obtained with the pivot method by fusing $S_{i-1,i}$ with $S_{i,i+1}^{\bar{I}_{i-1,i}}$. ©2004 by IEEE (Amigoni et al., 2004)

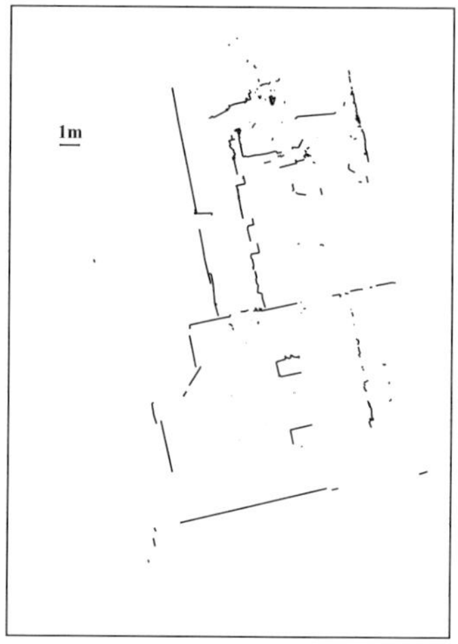

Figure 8. The final map obtained with the pivot method by fusing $S_{i-1,i}$ with $S_{i+1}^{\bar{I}_{i-1,i+1}}$. ©2004 by IEEE (Amigoni et al., 2004)

12.8 s. Furthermore, as already noted, when we integrate large-sized maps with many redundant spurious segments that represent the same part of the environment, the resulting maps are more noisy because of the error introduced when attempting to integrate maps with many overlapping segments.

Fig. 7 shows the final map, composed of 441 segments, obtained with the pivot method by fusing the partial map $S_{i-1,i}$ with $S_{i,i+1}^{\bar{t}_{i-1,i}}$. The map in Fig. 8 is composed of 358 segments and has been built by fusing the partial map $S_{i-1,i}$ with $S_{i+1}^{\bar{t}_{i-1,i+1}}$. This map presents fewer spurious segments and appears more "clean".

6. Conclusions

In this paper we have presented methods for matching pairs of scans composed of segments and for merging a sequence of partial maps in order to build a global map. In future research we aim at generalizing these methods to cases where the order in which the partial maps have to be integrated is not known. These generalized methods will provide an elegant solution to the problem of multirobot mapping since they will work when partial maps are acquired by a single robot at different times as well as when acquired by different robots in different locations.

Acknowledgments

The authors would like to thank Jean-Claude Latombe for his generous hospitality at Stanford University where this research was started, Héctor Gonzáles-Baños for sharing his programs and expertise with collecting laser range scan data, Paolo Mazzoni and Emanuele Ziglioli for the initial implementation of the fusion algorithm.

References

Amigoni, F., Gasparini, S., and Gini, M. (2004). Scan matching without odometry information. In *Proc. of the IEEE Int'l Conference on Robotics and Automation*, pages 3753–3758.

Burgard, W., Moors, M., and Schneider, F. (2002). Collaborative exploration of unknown environments with teams of mobile robots. In *Advances in Plan-Based Control of Robotic Agents*, pages 52–70. Springer-Verlag.

Fenwick, J. W., Newman, P. M., and Leonard, J. J. (2002). Cooperative concurrent mapping and localization. In *Proc. of the IEEE Int'l Conference on Robotics and Automation*, pages 1810–1817.

Gonzáles-Baños, H. H. and Latombe, J. C. (2002). Navigation strategies for exploring indoor environments. *Int'l Journal of Robotics Research*, 21(10-11):829–848.

Grimson, W. E. L. (1990). *Object recognition by computer: the role of geometric constraints*. The MIT Press.

Ko, J., Stewart, B., Fox, D., and Konolige, K. (2003). A practical, decision-theoretic approach to multi-robot mapping and exploration. In *Proc. of the IEEE/RSJ Int'l Conference on Intelligent Robots and Systems*, pages 3232–3238.

Konolige, K., Fox, D., Limketkai, B., Ko, J., and Stewart, B. (2003). Map merging for distributed robot navigation. In *Proc. of the IEEE/RSJ Int'l Conference on Intelligent Robots and Systems*.

Lu, F. and Milios, E. (1997). Robot pose estimation in unknown environments by matching 2D range scans. *Journal of Intelligent and Robotic Systems*, 18(3):249–275.

Martignoni III, A. and Smart, W. (2002). Localizing while mapping: A segment approach. In *Proc. of the Eighteen National Conference on Artificial Intelligence*, pages 959–960.

Simmons, R. G., Apfelbaum, D., Burgard, W., Fox, D., Moors, M., Thrun, S., and Younes, H. (2000). Coordination for multi-robot exploration and mapping. In *Proc. of the National Conference on Artificial Intelligence*, pages 852–858.

Thrun, S., Burgard, W., and Fox, D. (2000). A real-time algorithm for mobile robot mapping with applications to multi-robot and 3D mapping. In *Proc. of the IEEE Int'l Conference on Robotics and Automation*, pages 321–328.

Weiss, G., Wetzler, C., and Puttkamer, E. V. (1994). Keeping track of position and orientation of moving indoor systems by correlation of range-finder scans. In *Proc. of the IEEE/RSJ Int'l Conference on Intelligent Robots and Systems*, pages 12–16.

Zhang, L. and Ghosh, B. (2000). Line segment based map building and localization using 2D laser rangefinder. In *Proc. of the IEEE Int'l Conference on Robotics and Automation*, pages 2538–2543.

DISTRIBUTED COVERAGE OF UNKNOWN/UNSTRUCTURED ENVIRONMENTS BY MOBILE SENSOR NETWORKS

Ioannis Rekleitis
Currently at the Canadian Space Agency, Canada *
yiannis@cim.mcgill.ca

Ai Peng New
DSO National Laboratories, Singapore
naipeng@dso.org.sg

Howie Choset
Mechanical Engineering Department, Carnegie Mellon University, USA
choset@cmu.edu

Abstract In this paper we present an algorithmic solution for the distributed, complete coverage, path planning problem. Real world applications such as lawn mowing, chemical spill clean-up, and humanitarian de-mining can be automated by the employment of a team of autonomous mobile robots. Our approach builds on a single robot coverage algorithm. A greedy auction algorithm (a market based mechanism) is used for task reallocation among the robots. The robots are initially distributed through space and each robot is allocated a virtually bounded area to cover. Communication between the robots is available without any restrictions.

Keywords: Multi-Robot coverage, Automated De-mining, Market-based approach, Morse decomposition

*Work done while at Carnegie Mellon University.

L.E. Parker et al. (eds.),
Multi-Robot Systems. From Swarms to Intelligent Automata. Volume III, 145–155.

1. Introduction

The task of covering an unknown environment, common in many applications, is of high interest in a number of industries. Among them are manufacturers of automated vacuum/carpet cleaning machines and lawn mowers, emergency response teams such as chemical or radioactive spill detection and cleanup, and humanitarian de-mining. In addition, interesting theoretical problems have emerged especially in the areas of path planning, task (re)allocation and multi-robot cooperation.

The goal of complete coverage is to plan a path that would guide a robot to pass an end-effector (in our case equivalent to the footprint of the robot) over every accessible area of the targeted environment. In the single robot case, previous work has produced algorithms that guarantee complete coverage of an unknown arbitrary environment. Introducing multiple robots provides advantages in terms of efficiency and robustness but increases the algorithmic complexity.

Central in the multi-robot approach is the issue of communication. When communication is restricted to close proximity (Latimer-IV et al., 2002) or line of sight (Rekleitis et al., 2004) the robots have to remain together in order to avoid covering the same area multiple times. When unrestricted communication is available then the robots can disperse through the environment and proceed to cover different areas in parallel, constantly updating each other on their progress. The challenge in this case is to allocate regions to each robot such that no robot stays idle (thus all finish covering around the same time) and also to reduce the amount of time spent commuting among the different regions instead of covering. Providing an optimal solution for minimizing travel time is an NP-hard problem as it can be mapped into a multiple traveling salesman problem. An auction mechanism is used in order to re-allocate regions to be covered between robots in such a way that the path traveled between regions is reduced. The auction mechanism is a greedy heuristic based on the general market based approach.

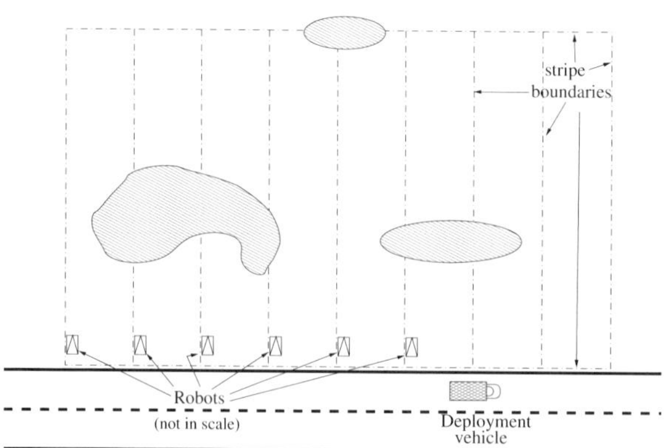

Figure 1 A large unknown area is divided up in vertical stripes. Each covering robot is assigned a stripe to cover. A deployment vehicle is utilized that distributes the robots at the beginning of the stripes. The robots do not know the layout at the interior of each stripe.

We assume that the robots know their position and orientation with respect to a global reference frame (e.g. via access to a GPS system). The robot sensors are able to detect both static obstacles and mobile robots, and differentiate between the two. The sensors have limited range and a good angular resolution.

The working paradigm in our approach is the application of humanitarian de-mining. A team of robots is deployed along one side of a field to be cleared, at regular intervals (as in Fig. 1). The interior of the field is unknown, partially covered with obstacles, and divided into a number of virtual stripes equal to the number of robots. Each robot is allocated initially the responsibility of the stripe it is placed at, and the coverage starts.

In the next section we present relevant background on the Coverage task and on the market based approach. Section 3 provides an overview of our algorithm and the next Section presents our experimental results in multiple simulated environments. Finally, Section 5 provides conclusions and future work.

2. Related Work

This work employs a single robot coverage algorithm for each individual robot and an auction mechanism to negotiate among robots which areas each robot would cover. Due to space limitations we will briefly outline the major approaches in multi-robot coverage (for a more detailed survey please refer to (Rekleitis et al., 2004)) and then we will discuss related work on market based mechanisms in mobile robotics. Finally, we present a brief overview of relevant terminology used in coverage and exact cellular decomposition. This work takes root in the Boustrophedon decomposition (Choset and Pignon, 1997), which is an exact cellular decomposition where each cell can be covered with simple back-and-forth motions.

Deterministic approaches have been used to cover specialized environments (Butler et al., 2001) sometimes resulting in repeat coverage (Latimer-IV et al., 2002, Kurabayashi et al., 1996, Min and Yin, 1998). Non-deterministic approaches include the use of neural networks (Luo and Yang, 2002), chemical traces (Wagner et al., 1999), and swarm intelligence (Ichikawa and Hara, 1999, Bruemmer et al., 2002, Batalin and Sukhatme, 2002). The non-deterministic approaches can not guarantee complete coverage.

2.1 Market-based Approach in Robotics

Cooperation and task allocation among mobile robots is crucial in multi-robot applications. To facilitate task re-allocation a new methodology based on market economy has gained popularity. For a comprehensive survey please refer to (Dias and Stentz, 2001). Currently market based approaches have been used to solve the multi-robot task allocation problem (Goldberg et al.,

2003) in the domains of: exploration (Berhault et al., 2003, Dias and Stentz, 2003), failure/malfunction detection and recovery (Dias et al., 2004), and box pushing (Gerkey and Mataric, 2002).

2.2 Boustrophedon/Morse Decomposition

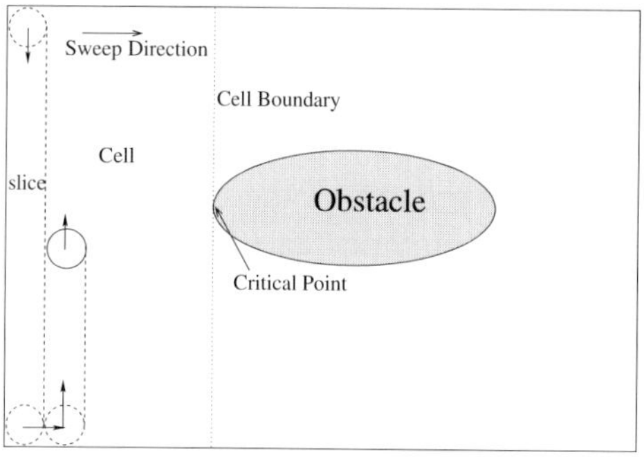

Figure 2 Illustrates the terms borrowed from single robot coverage with a single robot and one obstacle in the target environment. The robot is performing coverage with simple up-and-down motions.

To better describe the multi-robot coverage algorithm, we borrow the following terms from single robot coverage: *slice*, *cell*, *sweep direction*, and *critical point* (see Fig. 2). A *slice* is a subsection of a *cell* covered by a single, in our case vertical, motion. A *cell* is a region defined by the Boustrophedon decomposition where connectivity is constant. In our current work a cell is further constrained by the boundaries of the stripe (the space allocated to a robot). *Sweep direction* refers to the direction the slice is swept. Lastly, a *critical point* represents a point on an obstacle which causes a change in the cell connectivity. The critical points have been described in length in (Acar and Choset, 2000) (see Fig. 3a for an overview). We also borrow the concept of a Reeb graph, a graph representation of the target environment where the nodes are the critical points and the edges are the cells (Fig. 3b).

3. Algorithm Overview

Our approach consists of two behaviours, exploration and coverage. The robots initially try to trace the outline of the areas assigned to them in order to be more knowledgeable about the general layout of the free space. The connectivity of the free space is recorded in a graph that consists of the Reeb graph augmented with extra nodes (termed Steiner points) placed at the boundaries of the assigned stripes for each robot. The edges of the graph represent areas of accessible unexplored space and each edge belongs to a robot. During the exploration phase the robots exchange information and if the stripe a robot has

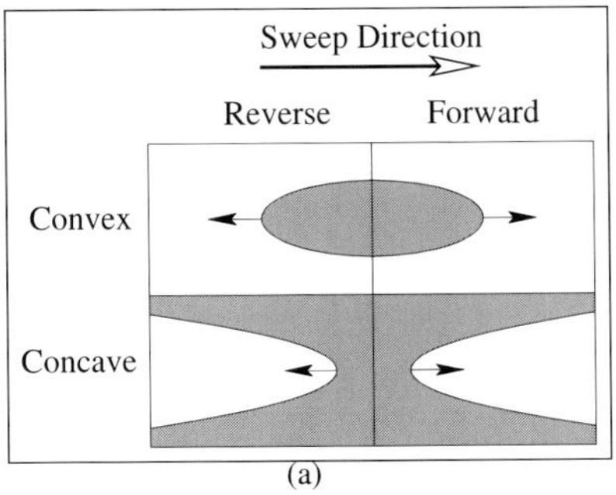

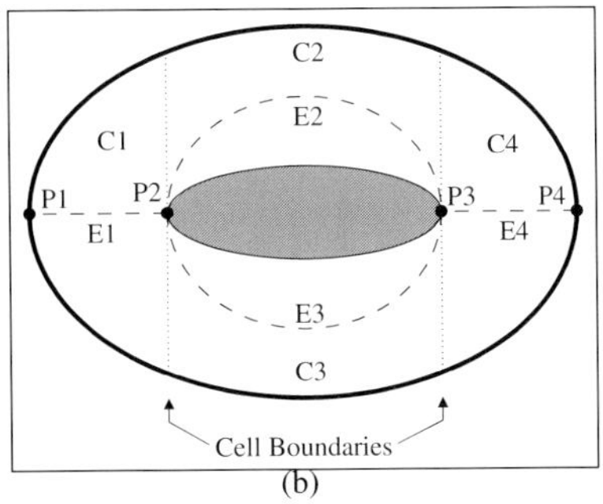

Figure 3. (a) Depicts the four types of critical points, based on concavity and the surface normal vector parallel to the sweep direction. Note that the shaded areas are obstacles and the arrows represent the normal vectors. (b) Here a simple Reeb graph is overlaid on top of a simple elliptical world with one obstacle. P1-P4 are critical points which represent graph nodes. E1-E4 represent edges which directly map to cells C1-C4.

assigned is not fully explored, then, that robot calls an auction for the task of exploring the remaining area of the stripe.

3.1 Cooperative Exploration

The robot uses the cycle algorithm developed in single robot Morse Decomposition for exploration of the stripe boundary. The cycle path is a simple closed path, i.e., by executing the cycle algorithm the robot always comes back to the point where it has started. This same cycle algorithm is used for both exploration and coverage. Before describing the cycle algorithm, we need to define 2 terms: lapping and wall following. Lapping is the motion along the slices while wall following is the motion along obstacle boundaries. A simple cycle algorithm execution will consist of forward lapping, forward wall following, reverse lapping and reverse wall following (as shown in Fig. 4a). This is sufficient for exploring the stripe boundary.

To explain the cooperative exploration algorithm, we will look at an example. Fig. 4b shows an unknown space with a single obstacle, being divided into 6 stripes. The Reeb graph of each robot is initialized with 2 critical points (**S**tart and **E**nd) and 5 Steiner points (representing the stripe boundaries).

The robots access their respective stripes and perform initial exploration using the cycle algorithm (forward lapping, forward wall following, reverse lapping and reverse wall following). During exploration, the robots modify their knowledge of the environment by updating the Reeb graph as they discover critical points and new information about the Steiner points. After completing a cycle, each robot shares its updated partial Reeb graph with the rest of the robots. At the end of the initial exploration, the updated global Reeb graph is as shown in Fig. 4c.

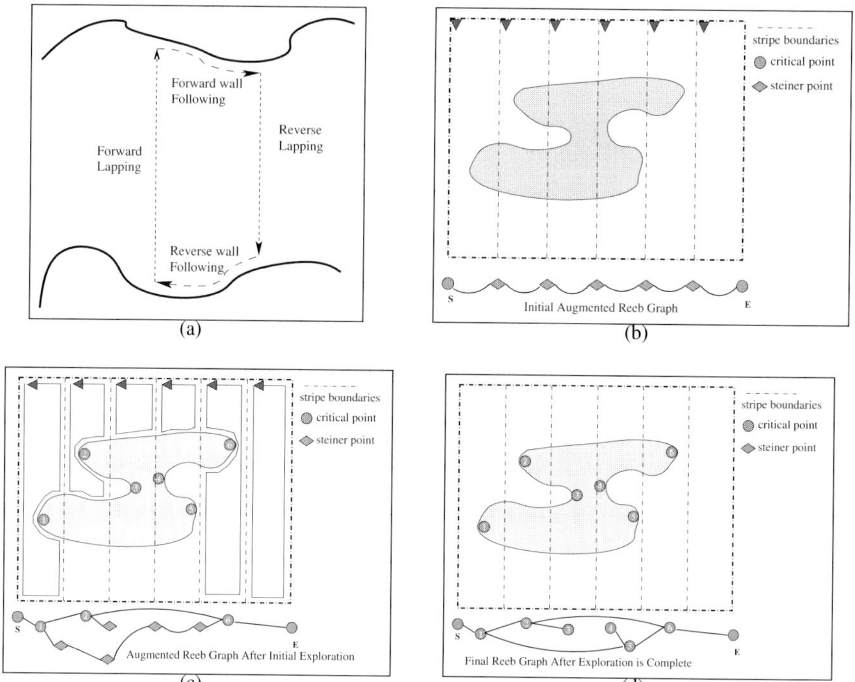

Figure 4. (a) A simple cycle path consisting of forward lapping, forward wall following, reverse lapping and reverse wall following. (b) Simple environment with initial Augmented Reeb Graph. (c) Initial exploration of stripes. (d) The final Reeb Graph after exploration is complete.

In the process of exploration, the robots will realize that there are spaces in their stripe that they are not able to reach easily. Those robots that are in such a situation will formulate the unreachable portions of the stripe as auction tasks and call auctions to re-allocate these parts of their stripe. In this manner, cooperative exploration is achieved. Fig. 4d shows the completed Reeb Graph after exploration is complete. Robots that do not have any exploration tasks can start performing partial coverage of known stripes in order not to waste time. Coverage of a cell is considered an atomic task, thus a robot that has started covering a cell would finish covering it before starting another task. The global Reeb graph is updated to represent the increased knowledge of the environment.

3.2 Cooperative Coverage

After all the stripe boundaries are completely explored (fully connected Reeb graph without Steiner points), the cells are owned by the robot that discovered them. The environment is fully represented by the Reeb graph, hence it is decomposed into a set of connected cells (the union of all the cells represents the free space), and all free space is allocated to the robots. Next the robots proceed to cover the cells under their charge. Coverage of a single cell is the same as single robot Morse Decomposition; if there are no obstacles within the cell, the coverage is a series of simple cycle paths. If there are obstacles within the cell, the robot performs incremental modification of the Reeb graph within that cell and shares the information with the other robots. If there is a robot that is without a task it calls an auction to offer its service to other robots. If all robots have completed their cell coverage and there are no uncovered cells in the Reeb graph, then the robots return to their starting positions and declare the environment covered.

3.3 Auctioning Tasks

A simple auction mechanism is used to investigate the feasibility of auction to enable cooperation among robots. At any auction a single task is auctioned out. In general, the auction mechanism operates as follows: (a) A robot discovers a new task and calls an auction with an initial estimated cost. (b) Other robots that are free to perform the task at a lower estimated cost, bid for the task. (c) When the auction time ends, the auctioneer selects the robot with the lowest bid and assigns the task. The winning robot adds the task into its task list and confirms that it accepts the task by sending an accept-task message back to the auctioneer. The auctioneer deletes the auction task and the task auction process concludes. As stated in the previous sections, auction is used in two separate ways: for cooperative exploration, and for cooperative coverage.

During exploration, a robot can encounter a situation where the stripe it is exploring is divided into two (or more) disconnected parts (see for example the middle stripe in Fig. 5a) because of an obstacle. The robot starts with forward lapping, encounters the obstacle and performs wall following. The wall following behaviour brings it to the stripe boundary associated with reverse lapping. As a result, the robot infers that there exists a disconnected stripe. At this point, it will formulate a new stripe to be explored and calls an auction for this new exploration task. Please note that the robots generally do not have sufficient information to know accurately the cost of performing the exploration task. It can only estimate the cost based on whatever information is available. Cost is the only parameter that decides the winning robot in an auction and it is thus the factor that determines the quality of cooperation. The estimation of the cost can be potentially a complex function of many variables (such as time spent, fuel expended, priorities of the task, capabilities of the robot). For this investigation, the task cost for the bidder is estimated based on 2 components: (a) Access cost: Based on the bidder's current estimated end point (the point where its currently executing atomic task will end), this is the shortest Manhattan distance to access the new stripe; (b) Exploration cost: Assuming that the robot can access the desired point in the stripe, this is the minimum distance that it needs to travel in order to explore the stripe completely (as parts of the stripe could already have been explored, the starting point of the exploration could result in different costs for different robots).

When an initial estimate of the cells is available (exploration is complete) the robot that has discovered a cell is initially responsible for covering it. The robot without any tasks will offer its service by also calling an auction. Any robot that has extra cells (less the cell that it is currently covering) will offer one of the cells, based on the auctioneer's position. Each robot without extra cells will estimate the current cell workload and offer to share its cell coverage task if it is greater than a threshold. The auctioneer prefers to takeover a cell rather than to share coverage of a cell. It will use the estimated distance to access the cell as a selection criteria if there are more than one cell on offer.

4. Experimental Results

The distributed coverage algorithm was implemented in simulation using Player and Stage (Gerkey et al., 2001) with 3 robots. We adopted a highly distributed system architecture because it can quickly respond to problems involving one (or a few) robots, and is more robust to point failures and the changing dynamics of the system. Our architecture is based on the layered approach that has been used for many single-agent autonomous systems (Schreckenghost et al., 1998, Wagner et al., 2001). We are employing two layers for each robot instead of the traditional three layers: Planning and Behaviour. The *upper* layer consists of *Planner* and *Model* and the lower layer is *Behaviour*. Model is where the Reeb graph resides. Planner is where Morse Decomposition, auction mechanism, task scheduling and task monitoring take place. The Behaviour process serves the same function as in traditional layered architecture, controlling the robots to perform atomic tasks such as Goto, Follow Wall and Lapping.

A sample environment for testing the algorithm is shown in Fig. 5a. Each robot is allocated a stripe and the Planner of each robot receives the stripe information. The Planner determines the point where it wants to access the stripe and sends the way-point to the Behaviour process for execution. After accessing the stripe, the Behaviour process sends a message to the Planner informing the Planner that access of the stripe is completed. Based on the stripe information and the robot pose, the Planner plans for Forward Lapping and sends this task to the Behaviour. The Behaviour executes the forward lapping task. For this task, the 3 robots experience different terminating conditions because of the environment: The left and the right robots complete the exploration of their stripes without any problems. The middle robot realizes that it can not complete the exploration of its stripe and calls an auction. The robot on the

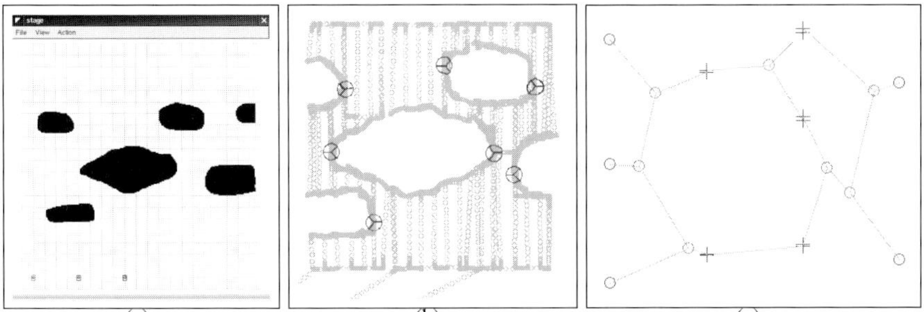

(a) (b) (c)

Figure 5. (a) The environment and the three robots at the starting position in Stage. (b) The traces of the robots (marked as circles which are smaller than the footprint) and the critical points encountered. (c) The augmented Reeb graph with the critical points (circles) and the Steiner points (crosses).

right wins the auction and proceeds to explore the remaining part of the middle stripe. In the mean time the left and middle robots start partial coverage. Finally when exploration is complete the robots exchange cells via auction and completely cover the environment. Fig. 5c shows the Reeb graph after exploration is completed. Fig. 5b shows the trace of the three robots plotted as circles (the trace is smaller than the robot footprint for illustration purposes).

During our experiments the robots continuously explored and covered the environment. After a few auctions it was impossible to predict which task was scheduled next by each robot. It is worth noting though that the distance traveled by each robot was approximately the same thus showing that the workload was distributed evenly.

5. Summary

In this paper we presented an algorithmic approach to the distributed, complete coverage, path planning problem. Under the assumption of global communication among the robots, each robot is allocated an area of the unknown environment to cover. An auction mechanism is employed in order to facilitate cooperative behaviour among the robots and thus improve their performance. In our approach no robot remains idle while there are areas to be covered.

For future work, we would like to compare the performance between the distributed approach described here with the formation-based approach with limited communication presented in (Rekleitis et al., 2004). Augmenting the cost function to take into account individual robot capabilities (especially in heterogeneous teams) is an important extension. Accurate localization is a major challenge in mobile robotics; we would like to take advantage of the meeting of the robots in order to improve the localization quality via cooperative localization (Roumeliotis and Rekleitis, 2004). Finally, developing more accurate cost estimates for the different tasks is one of the immediate objectives.

Acknowledgments

The authors wish to thank Edward Rankin for his help with the algorithm implementation and experimentation; Vincent Lee-Shue and Sam Sonne for providing valuable input during the early stages of this project; Bernadine Dias, Danni Goldberg, Rob Zlot and Marc Zink for their help with the Market based approach. Finally, Luiza Solomon provided valuable insights on the software design and an implementation of a graph class. Furthermore, we would like to acknowledge the generous support of the DSO National Laboratories, Singapore, the Office of Naval Research, and the National Science Foundation.

References

Acar, E. U. and Choset, H. (2000). Critical point sensing in unknown environments. In *Proc. of the IEEE International Conference on Robotics & Automation.*

Batalin, M. A. and Sukhatme, G. S. (2002). Spreading out: A local approach to multi-robot coverage. In *6th International Symposium on Distributed Autonomous Robotics Systems*, Fukuoka, Japan.

Berhault, M., Huang, H., Keskinocak, P., Koenig, S., Elmaghraby, W., Griffin, P., and Kleywegt, A. (2003). Robot exploration with combinatorial auctions. In *IEEE/RSJ Int. Conference on Intelligent Robots and Systems*, volume 2, pages 1957–1962.

Bruemmer, D. J., Dudenhoeffer, D. D., Anderson, M. O., and McKay, M. D. (2002). A robotic swarm for spill finding and perimeter formation. *Spectrum.*

Butler, Z., Rizzi, A., and Hollis, R. (2001). Distributed coverage of rectilinear environments. In *Proc. of the Workshop on the Algorithmic Foundations of Robotics.*

Choset, H. and Pignon, P. (1997). Coverage path planning: The boustrophedon cellular decomposition. In *International Conference on Field and Service Robotics*, Canberra, Australia.

Dias, M. B. and Stentz, A. T. (2001). A market approach to multirobot coordination. Technical Report CMU-RI -TR-01-26, Robotics Institute, Carnegie Mellon University, Pittsburgh, PA.

Dias, M. B. and Stentz, A. T. (2003). Traderbots: A market-based approach for resource, role, and task allocation in multirobot coordination. Technical Report CMU-RI -TR-03-19, Robotics Institute, Carnegie Mellon University, Pittsburgh, PA.

Dias, M. B., Zinck, M., Zlot, R., and Stentz, A. T. (2004). Robust multirobot coordination in dynamic environments. In *International Conference on Robotics & Automation*, pages 3435–3442, New Orleans, LA.

Gerkey, B. and Mataric, M. (2002). Sold!: auction methods for multirobot coordination. *IEEE Transactions on Robotics and Automation*, 18(5):758 – 768.

Gerkey, B. P., Vaughan, R. T., StÃÿy, K., Howard, A., Sukhatme, G. S., and Mataric, M. J. (2001). Most valuable player: A robot device server for distributed control. In *IEEE/RSJ International Conference on Intelligent Robots and Systems (IROS 2001)*, pages 1226–1231, Wailea, Hawaii.

Goldberg, D., Cicirello, V., Dias, M. B., Simmons, R., Smith, S. F., and Stenz, A. (2003). Market-based multi-robot planning in a distributed layered architecture. In *Multi-Robot Sys-*

tems: From Swarms to Intelligent Automata: Proceedings from the 2003 International Workshop on Multi-Robot Systems, volume 2, pages 27–38. Kluwer Academic Publishers.

Ichikawa, S. and Hara, F. (1999). Characteristics of object-searching and object-fetching behaviors of multi-robot system using local communication. In *IEEE International Conference on Systems, Man, and Cybernetics, (IEEE SMC '99)*, volume 4, pages 775 –781.

Kurabayashi, D., Ota, J., Arai, T., and Yoshida, E. (1996). Cooperative sweeping by multiple mobile robots. In *1996 IEEE International Conference on Robotics and Automation*, volume 2, pages 1744 –1749.

Latimer-IV, D., Srinivasa, S., Lee-Shue, V., Sonne, S. S., Choset, H., and Hurst, A. (2002). Toward sensor based coverage with robot teams. In *Proc. 2002 IEEE International Conference on Robotics & Automation,*.

Luo, C. and Yang, S. (2002). A real-time cooperative sweeping strategy for multiple cleaning robots. In *IEEE Internatinal Symposium on Intelligent Control*, pages 660 –665.

Min, T. W. and Yin, H. K. (1998). A decentralized approach for cooperative sweeping by multiple mobile robots. In *1998 IEEE/RSJ International Conference on Intelligent Robots and Systems*, volume 1, pages 380–385.

Rekleitis, I., Lee-Shue, V., New, A. P., and Choset, H. (2004). Limited communication, multi-robot team based coverage. In *IEEE International Conference on Robotics and Automation*, pages 3462–3468, New Orleans, LA.

Roumeliotis, S. I. and Rekleitis, I. M. (2004). Propagation of uncertainty in cooperative multi-robot localization: Analysis and experimental results. *Autonomous Robots*, 17(1):41–54.

Schreckenghost, D., Bonasso, P., Kortenkamp, D., and Ryan, D. (1998). Three tier architecture for controlling space life support systems. In *IEEE Int. Joint Symposia on Intelligence and Systems*, pages 195 – 201.

Wagner, I., Lindenbaum, M., and Bruckstein, A. (1999). Distributed covering by ant-robots using evaporating traces. *IEEE Transactions on Robotics and Automation*, 15(5):918–933.

Wagner, M., Apostolopoulos, D., Shillcutt, K., Shamah, B., Simmons, R., and Whittaker, W. (2001). The science autonomy system of the nomad robot. In *IEEE International Conference on Robotics and Automation*, volume 2, pages 1742 – 1749.

V

MOTION PLANNING AND CONTROL

REAL-TIME MULTI-ROBOT MOTION PLANNING WITH SAFE DYNAMICS *

James Bruce and Manuela Veloso
Computer Science Department
Carnegie Mellon University
Pittsburgh PA 15213, USA
{jbruce,mmv}@cs.cmu.edu

Abstract This paper introduces a motion planning system for real-time control of multiple high performance robots in dynamic and unpredictable domains. It consists of a randomized realtime path planner, a bounded acceleration motion control system, and a randomized velocity-space search for collision avoidance of multiple moving robots. The realtime planner ignores dynamics, simplifying planning, while the motion control ignores obstacles, allowing a closed form solution. This allows up to five robots to be controlled 60 times per second, but collisions can arise due to dynamics. Thus a randomized search is performed in the robot's velocity space to find a safe action which satisfies both obstacle and dynamics constraints. The system has been fully implemented, and empirical results are presented.

Keywords: realtime path planning, multirobot navigation

1. Introduction

All mobile robots share the need to navigate, creating the problem of motion planning. When multiple robots are involved, the environment becomes dynamic, and when noise or external agents are present, the environment also becomes unpredictable. Thus the motion planning system must be able to cope with dynamic, unpredictable domains. To take advantage of high performance robots, and respond quickly to external changes in the domain, the system must also run at real-time rates. Finally, navigation at high speeds means respecting dynamics constraints in the robot motion to avoid collisions while staying

*This work was supported by United States Department of the Interior under Grant No. NBCH-1040007, and by Rockwell Scientific Co., LLC under subcontract No. B4U528968 and prime contract No. W911W6-04-C-0058 with the US Army. The views and conclusions contained herein are those of the authors, and do not necessarily reflect the position or policy of the sponsoring institutions, and no official endorsement should be inferred.

L.E. Parker et al. (eds.),
Multi-Robot Systems. From Swarms to Intelligent Automata. Volume III, 159–170.

within the operational bounds of the robot. When multiple robots are introduced, the system must find solutions where no robots collide while satisfying each robot's motion constraints. This paper describes a domain with these properties, and a motion planning system which satisfies the requirements for it. In the remainder of this section, the domain will be presented, and following sections will describe the three major parts of the system mentioned above. The system will then be evaluated in its entirety as to how well it solves navigation tasks.

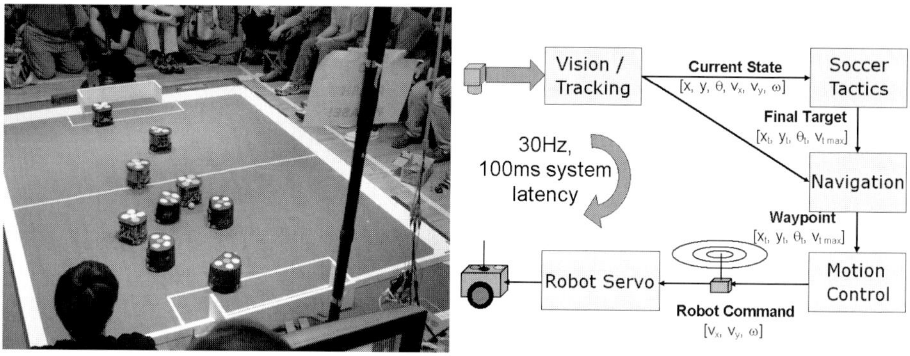

Figure 1. Two teams are shown playing soccer in the RoboCup small size league (left), and the the overall system architecture for CMDragons (right).

The motivating domain for this work is the RoboCup F180 "small size" league (Kitano et al., 1995). It involves teams of five small robots, each up to 18cm in diameter and 15cm height. The robot teams are entered into a competition to play soccer against opponent teams fielded by other research groups. During the game no human input is allowed, thus the two robot teams must compete using full autonomy in every aspect of the system. The field of play is a carpet measuring 4.9m by 3.8m, with a 30cm border around the field for positioning outside the field of play (such as for free kicks). An earlier (half size) version of the field is pictured on the left of Figure 1. Offboard communication and computation is allowed, leading nearly every team to use a centralized approach for most of the robot control. The data flow in our system, typical of most teams, is shown on the right of Figure 1 (Bruce et al., 2003). Sensing is provided by two or more overhead cameras, feeding into a central computer to process the image and locate the 10 robots and the ball on the field 30-60 times per second. These locations are fed into an extended Kalman filter for tracking and velocity estimation, and then sent to a "soccer" module which implements the team strategy using various techniques. The soccer system implements most of its actions through a navigation module, which provides path planning, obstacle avoidance, and motion control. Finally, velocity commands are sent to the robots via a serial radio link. Due to its competitive nature, teams have pushed robotic technology to its limits, with the small robots travelling over $2m/s$, accelerations between $3-6m/s^2$, and kicking the golf ball used in the game at $4-6m/s$. Their speeds require every module to run in realtime to minimize latency, while leaving enough computing resources for the other modules to operate. In addition, the algorithms must operate robustly due to the full autonomy requirement.

Navigation is a critical component in the overall system described above, and the one we will focus on in this paper. The system described here is meant as a drop-in replacement for the navigation module used successfully by our team since 2002, and building on experience gained since 1997 working on fast navigation for small high performance robots (Bowling and Veloso, 1999). For our model, we will assume a centralized system, although the interaction required between robots will be minimized which should make decentralization a straightforward extension. It comprises of three critical parts; A path planner, a motion control system, and a velocity space search for safe motions. Although motivated by the specific requirements of the RoboCup domain, it should be of general applicability to domains where several robots must navigate within a closed space where both high performance and safety are desired.

2. Path Planning

For path planning, the navigation system models the environment in 2D, and also ignoring dynamics constraints, which will instead be handled by a later module. The algorithm used is the ERRT extension of the RRT-GoalBias planner (LaValle, 1998, LaValle and James J. Kuffner, 2001). Due to the speed of the algorithm, a new plan can be constructed each control cycle, allowing it to track changes in the dynamic environment without the need for replanning heuristics. A more thorough description of our previous work on ERRT can be found in (Bruce and Veloso, 2002). Since then, a new more efficient implementation has been completed, but the underlying algorithm is the same. It is described here in enough detail to be understood for the evaluations later in the paper.

Rapidly-exploring random trees (RRTs) (LaValle, 1998) employ randomization to explore large state spaces efficiently, and form the basis for a family of probabilistically complete, though non-optimal, kinodynamic path planners (LaValle and James J. Kuffner, 2001). Their strength lies in that they can efficiently find plans in relatively open or high dimensional spaces because they avoid the state explosion that discretization faces. A basic planning algorithm using RRTs is shown in Figure 2, and the steps are as follows: Start with a trivial tree consisting only of the initial configuration. Then iterate: With probability p, find the nearest point in the current tree and extend it toward the goal g. Extending means adding a new point to the tree that extends from a point in the tree toward g while maintaining whatever motion constraints exist. Alternatively, with probability $1-p$, pick a point x uniformly from the configuration space, find the nearest point in the current tree, and extend it toward x. Thus the tree is built up with a combination of random exploration and biased motion towards the goal configuration.

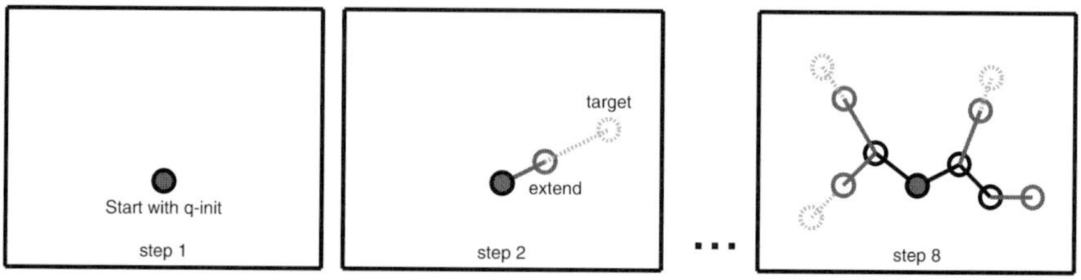

Figure 2. Example growth of an RRT tree for several steps. Each iteration, a random target is chosen and the closest node in the tree is "extended" toward the target, adding another node to the tree.

To convert the RRT algorithm into a planner, we need two simple additions. One is to restrict the tree to free space, where it will not collide with obstacles. This can be accomplished by only adding nodes for extensions that will not hit obstacles. To make the tree into a planner, we only need to stop once the tree has reached a point sufficiently close to the goal location. Because the root of the tree is the initial position of the robot, tracing up from any leaf gives a valid path through free space between that leaf and the initial position. Thus finding a leaf near the goal is sufficient to solve the planning problem.

Execution Extended RRT (ERRT) adds the notion of a waypoint cache, which is a fixed-size lossy store of nodes from successful plans in previous iterations of the planner. Whenever a plan is found, all nodes along the path are added to the cache with random replacement of previous entries. Then during planning, random targets are now chosen from three sources instead of two. In other words, with probability p it picks the goal, with probability q it picks a random state from the waypoint cache, and with the remaining $1-p-q$ it picks a random state in the environment.

In order to implement ERRT we need an *extend* operator, a distance function between robot states, a distribution for generating random states in the environment, and a way of determining the closest point in a tree to a given target state. Our implementation uses Euclidean distance for the distance function and the uniform distribution for generating random states. The nearest state in the tree is determined using KD-Trees, a common technique for speeding up nearest neighbor queries. Finally the *extend* operator it simply steps a fixed distance along the path from the current state to the target. For a planner ignoring dynamics, this is the simplest way to guarantee the new state returned is closer to the intermediate target than the parent. Our step size is set to the minimum of the robot's radius and the distance to the randomly chosen target. An image of the planner running in simulation is shown in Figure 3, and a photograph of a real robot controlled by the planner is shown in Figure 4. To simplify input to the motion control, the resulting plan is reduced to a single target point, which is the furthest node along the path that can be reached with a straight line that does not hit obstacles. This simple postprocess smooths out local non-optimalities in the generated plan.

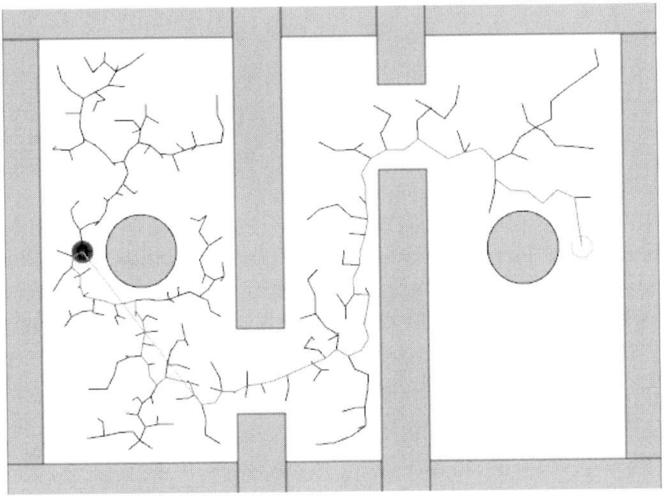

Figure 3. A robot on the left finds a path to a goal on the right using ERRT.

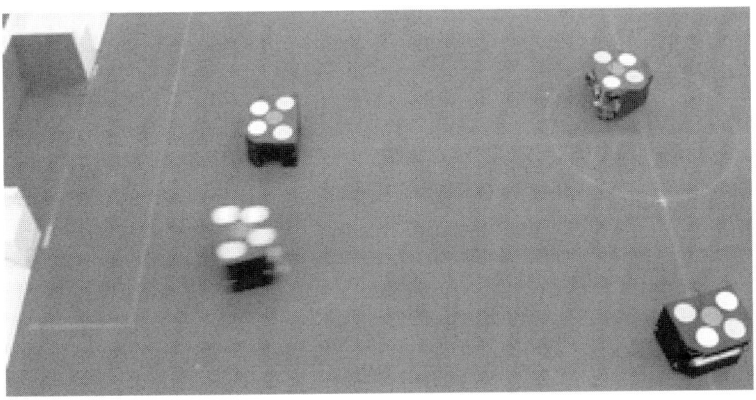

Figure 4. A robot (lower left) navigating at high speed through a field of static obstacles.

3. Motion Control

Once the planner determines a waypoint for the robot to drive to in order to move toward the goal, this target state is fed to the motion control layer.

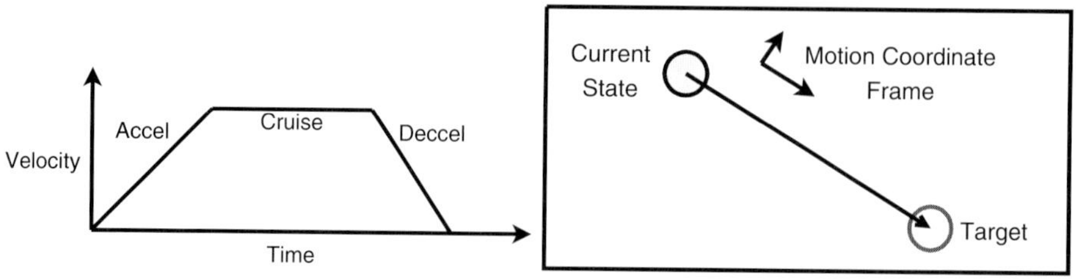

Figure 5. Our motion control approach uses trapezoidal velocity profiles. For the 2D case, the problem can decomposed into two 1D problems, one along the difference between the current state and the target state, and the other along its perpendicular.

The motion control system is responsible for commanding the robot to reach the target waypoint from its current state, while subject to the physical constraints of the robot. The model we will take for our robot is a three or four wheeled omnidirectional robot, with bounded acceleration and a maximum velocity. The acceleration is bounded by a constant on two independent axes, which models a four-wheeled omnidirectional robot well. In addition, deceleration is a separate constant from acceleration, since braking can often be done more quickly than increasing speed. The approach taken for motion control is the well known trapezoidal velocity profile. In other words, to move along a dimension, the velocity is increased at maximum acceleration until the robot reaches its maximum speed, and then it decelerates at the maximum allowed value to stop at the final destination (Figure 5. The area traced out by the trapezoid is the displacement effected by the robot. For motion in 2D, the problem is decomposed as a 1D motion problem along the axis from the robots' current position to the desired target, and another 1D deceleration perpendicular to that axis.

While the technique is well known, the implementation focuses on robustness even in the presence of numerical inaccuracies, changing velocity or acceleration constraints, and the inability to send more than one velocity command per cycle. First, for stability in the 2D case, if the initial and target points are close, the coordinate frame becomes degenerate. In that case the last coordinate frame above the distance threshold is used. For the 1D case, the entire velocity profile is constructed before calculating the command, so the behavior over the entire command period ($1/60$ to $1/30$ of a second) can be represented. The calculation proceeds in the following stages:

- If the current velocity is opposite the difference in positions, decelerate to a complete stop
- Alternatively, if the current velocity will overshoot the target, decelerate to a complete stop
- If the current velocity exceeds the maximum, decelerate to the maximum

- Calculate a triangular velocity profile that will close the gap
- If the peak of the triangular profile exceeds the maximum speed, calculate a trapezoidal velocity profile

Although these rules construct a velocity profile that will reach the target point if nothing impedes the robot, limited bandwidth to the robot servo loop necessitates turning the full profile into a single command for each cycle. The most stable version of generating a command was to simply select the velocity in the profile after one command period has elapsed. Using this method prevents overshoot, but does mean that very small short motions will not be taken (when the entire profile is shorter than a command period). In these cases it may be desirable to switch to a position based servo loop rather than a velocity base servo loop if accurate tracking is desired.

4. Velocity Space Safety Search

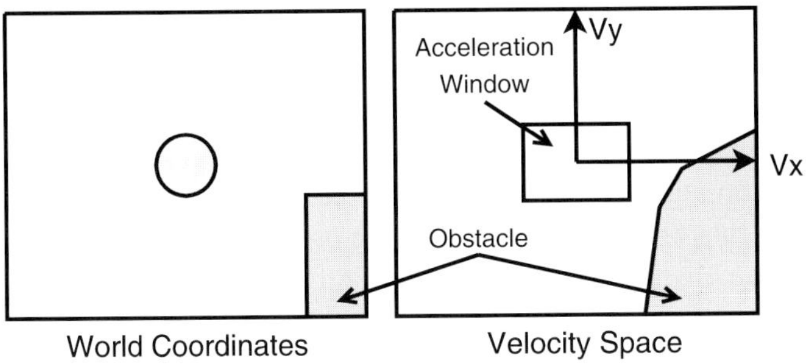

Figure 6. Example environment shown in world space and velocity space.

In the two previous stages in the overall system, the planner ignored dynamics while the motion control ignored obstacles, which has no safety guarantees in preventing collisions between the agent and the world, or between agents. The "Dynamic Window" approach (Fox et al., 1997) is a search method which elegantly solves the first problem of collisions between the robotic agent and the environment. It is a local method, in that only the next velocity command is determined, however it can incorporate non-holonomic constraints, limited accelerations, maximum velocity, and the presence of obstacles into that determination, thus guaranteeing safe motion for a robot. The search space is the velocities of the robot's actuated degrees of freedom. The two developed cases are for synchro-drive robots with a linear velocity and an angular velocity, and for holonomic robots with two linear velocities (Fox et al., 1997, Brock and Khatib, 1999). In both cases, a grid is created for the velocity space, reflecting an evaluation of velocities falling in each cell. First, the obstacles of the environment are considered, by assuming the robot travels at a cell's velocity for one control cycle and then attempts to brake at maximum deceleration while following that same trajectory. If the robot cannot come to a stop before hitting an obstacle along that trajectory, the cell is given an evaluation of zero. Next, due to limited accelerations, velocities are limited to a small window that can be reached within the acceleration limits over the next control cycle (for a holonomic robot this is a rectangle around the current velocities). An example is shown in Figure 6. Finally, the remaining valid velocities are scored using a heuristic distance to the goal. It was used successfully in robots moving up to 1m/s in cluttered office environments with dynamically placed obstacles (Brock and Khatib, 1999).

Our approach first replaces the grid with randomized sampling, with memory of the acceleration chosen in the last frame. Static obstacles are handled in much the same way, by checking if braking at maximum deceleration will avoid hitting obstacles. The ranking of safe velocities is done by choosing the one with the minimum Euclidean distance to the desired velocity given by the motion control. For moving bodies (robots, in this case) we have to measure the minimum distance between two accelerating bodies. The distance can be computed by solving two fourth degree polynomials (one while both robots are decelerating, and the second while one robot has stopped and the other is still trying to stop). For simplicity however we solved the polynomials numerically. Using this primitive, the velocity calculations proceeded as follows. First, all the commands were initialized to be maximum deceleration for stopping. Then as the robots chose velocities in order, they consider only velocities that don't hit an environment obstacle or hit the other robots based on the latest velocity command they have chosen. The first velocity tried is the exact one calculated by motion control. If that velocity is valid (which it normally is), then the sampling stage can be skipped. If it is not found, first the best solution from the last cycle is tried, followed by uniform sampling of accelerations for the next frame. Because the velocities are chosen for one cycle, followed by maximum deceleration, in the next cycle the robots should be safe to default to their maximum deceleration commands. Since this "chaining" assumption always guarantees that deceleration is an option, robots should always be able to stop safely. In reality, numerical issues, the limitations of sampling, and the presence of noise make this perfect-world assumption fail. To deal with this, we can add small margins around the robot, and also rank invalid commands. In the case where no safe velocity can be found, the velocity chosen is the one which minimizes the interpenetration depth with other robots or obstacles. This approach often prevents the robot from hitting anything at all, though it depends on the chance that the other robots involved *can* choose commands to avoid the collision.

Taken together, the three parts of the navigation system consisting of planner, motion control, and safety search, solve the navigation problem in a highly factored way. That is, they depend only on the current state and expected next command of the other robots. Factored solutions can be preferable to global solutions including all robots degrees of freedom as one planning problem for two reasons. One is that factored solutions tend to be much more efficient, and the second is that they also have bounded communication requirements if the algorithm needs to be distributed among agents. Each robot must know the position and velocity of the other robots, and communicate any command it decides, but it does not need to coordinate at any higher level beyond that.

5. Evaluation and Results

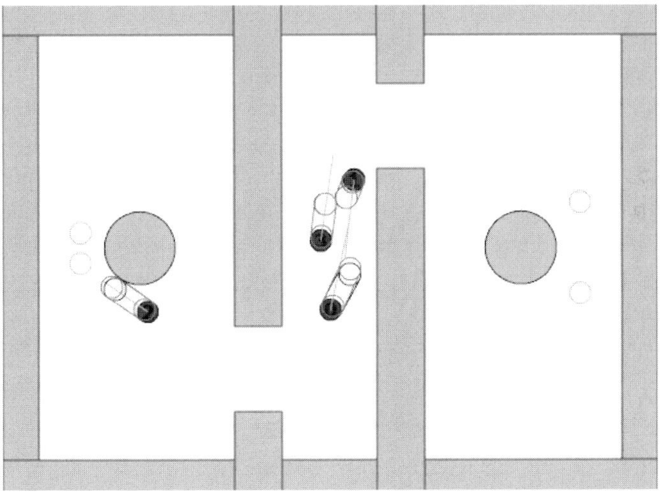

Figure 7. Multiple robots navigating traversals in parallel. The outlined circles and lines extending from the robots represent the chosen command followed by a maximum rate stop.

The evaluation environment the same as shown in Figure 3, which matches the size of the RoboCup small size field, but has additional large obstacles with narrow passages in order to increase the likelihood of robot interactions. The $90mm$ radius robots represent the highest performance robots we have used in RoboCup. Each has a command cycle of $1/60$ sec, a maximum velocity of $2m/s$, acceleration of $3m/s^2$, and deceleration of $6m/s^2$. For tests, four robots were given the task of traveling from left to right and back again four times, with each robot having separate goal points separated in the y axis. Because different robots have different path lengths to travel, after a two traversals robots start interacting while trying to move in opposed directions. Figure 7 shows an example situation. The four full traversals of all the robots took about 30 seconds of simulated time.

For the evaluation metric, we chose interpenetration depth multiplied by the time spent in those invalid states. To more closely model a real system, varying amounts of position sensor error were added, so that the robot's reported position was a Gaussian deviate of its actual position. This additive random noise represents vision error from overhead tracking systems. Velocity sensing and action error were not modelled here for simplicity; These errors depend heavily on the specifics of the robot and lack a widely applicable model. First, we compared using planner and motion control but enabling or disabling the safety search. Each data point is the average of 40 runs, representing about 20 minutes of simulated run time. Figure 8 shows the results, where it is clearly evident that the safety search significantly decreases the total interpenetration time. Without the safety search, increasing the vision error makes little difference in the length and depth of collisions. Next, we evaluated the safety search using different margins of $1 - 4mm$ around the $90mm$ robots, plotted against increasing vision error (Figure 9. As one would expect, with little or no vision error even small margins suffice for no collisions, but as the error increases there is a benefit to higher margins for the safety search, reflecting the uncertainty in the actual position of the robot. As far as running times, the realtime operation of ERRT has been maintained, with a mean time of execution is $0.70ms$ without the velocity safety search and $0.76ms$ with it. These reflect the fact that the planning problem is usually easy, and interaction with other robots that require sampling are rare. Focusing on the longer times, the 95% percentiles are $1.96ms$ and $2.04ms$, respectively.

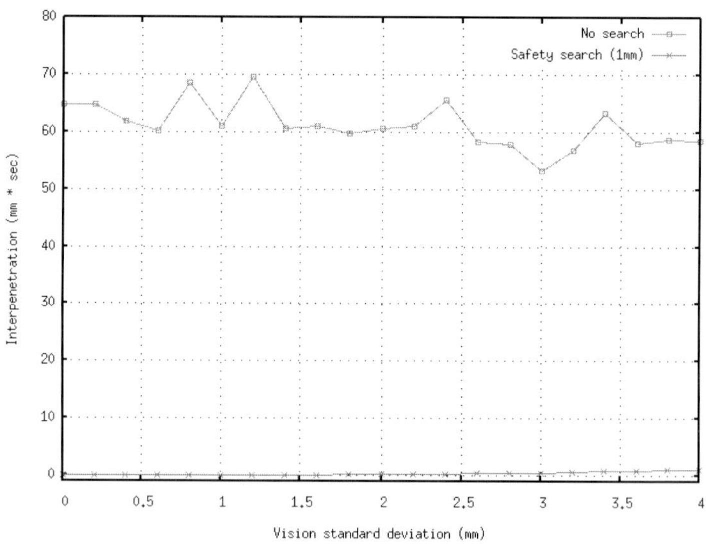

Figure 8. Comparison of navigation with and without safety search. Safety search significantly decreases the metric of interpenetration depth multiplied by time of interpenetration.

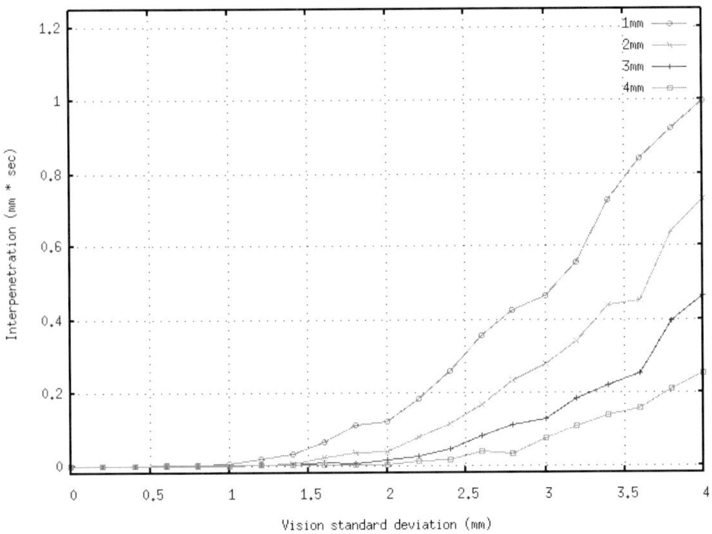

Figure 9. Comparison of several margins under increasing vision error.

6. Conclusion

This paper described a navigation system for the real-time control of multiple high performance robots in dynamic domains. Specifically, it addressed the issue of multiple robots operating safely without collisions in a domain with acceleration and velocity constraints. While the current solution is centralized, it does not rely on complex interactions and would rely on minimal communication of relative positions, velocities, and actions to distribute the algorithm. Like most navigation systems however, it makes demands of sensor accuracy that may not yet be practical for multiple distributed robots with only local sensing. Its primary contribution is to serve as an example and model of a complete system for similar problem domains.

References

Bowling, M. and Veloso, M. (1999). Motion control in dynamic multi-robot environments. In *International Symposium on Computational Intelligence in Robotics and Automation (CIRA'99).*

Brock, O. and Khatib, O. (1999). High-speed navigation using the global dynamic window approach. In *Proceedings of the IEEE International Conference on Robotics and Automation.*

Bruce, J., Bowling, M., Browning, B., and Veloso, M. (2003). Multi-robot team response to a multi-robot opponent team. In *Proceedings of the IEEE International Conference on Robotics and Automation.*

Bruce, J. and Veloso, M. (2002). Real-time randomized path planning for robot navigation. In *Proceedings of the IEEE Conference on Intelligent Robots and Systems (IROS).*

Fox, D., Burgard, W., and Thrun, S. (1997). The dynamic window approach to collision avoidance. *IEEE Robotics and Automation Magazine*, 4.

Kitano, H., Asada, M., Kuniyoshi, Y., Noda, I., and Osawa, E. (1995). Robocup: The robot world cup initiative. In *Proceedings of the IJCAI-95 Workshop on Entertainment and AI/ALife*.

LaValle, S. M. (1998). Rapidly-exploring random trees: A new tool for path planning. In *Technical Report No. 98-11*.

LaValle, S. M. and James J. Kuffner, J. (2001). Randomized kinodynamic planning. In *International Journal of Robotics Research, Vol. 20, No. 5*, pages 378–400.

A MULTI-ROBOT TESTBED FOR BIOLOGICALLY-INSPIRED COOPERATIVE CONTROL

Rafael Fierro and Justin Clark
MARHES *Laboratory*
Oklahoma State University
Stillwater, OK 74078, USA
{rafael.fierro, justin.clark} @okstate.edu

Dean Hougen and Sesh Commuri
REAL *Laboratory*
University of Oklahoma
Norman, OK 73072, USA
{hougen, scommuri} @ou.edu

Abstract In this paper a multi-robot experimental testbed is described. Currently, the testbed consists of five autonomous ground vehicles and two aerial vehicles that are used for testing multi-robot cooperative control algorithms. Each platform has communication, on-board sensing, and computing. Robots have *plug-and-play* sensor capability and use the Controller Area Network (CAN) protocol, which maintains communication between modules. A biologically-inspired cooperative hybrid system is presented in which a group of mobile robotic sensors with limited communication are able to search for, locate, and track a perimeter while avoiding collisions.

Keywords: Multi-robot cooperation, perimeter detection, biologically inspired hybrid system.

1. Introduction

Control problems for teams of robots can have applications in a wide variety of fields. At one end of the spectrum are multi-vehicle systems that are not working together, but need to be coordinated to the extent that they do not interfere with one another. Typical applications in this area are in air-traffic control. At the other end of the spectrum are teams of robots operating cooperatively and moving in formations with precisely defined geometries that need

L.E. Parker et al. (eds.),
Multi-Robot Systems. From Swarms to Intelligent Automata. Volume III, 171–182.

to react to the environment to achieve stated goals (Das et al., 2002). In the middle ground there are groups of vehicles that have some common goals, but may not have precisely defined desired trajectories. One control problem in this area is robots with behavior that is analogous to the schooling or herding behavior of animals (Jadbabaie et al., 2003).

The need for unmanned ground vehicles (UGVs) capable of working together as a sensing network for both industrial and military applications has led to the design of the multi-vehicle MARHES testbed. The testbed consists of five autonomous ground vehicles and two UAV's that are used for testing multi-agent cooperative control algorithms in applications such as perimeter detection, search and rescue, and battlefield assessment to mention just a few. The main goals when designing this testbed were to make it affordable, modular, and reliable. This balance of price and performance has not been available in the commercial market yet. The *plug-and-play* sensor modules allow modularity. Reliability is achieved with the Controller Area Network (CAN) which maintains data integrity.

The rest of the chapter is organized as follows. The multi-robot testbed is described in section 2. Section 3 presents a biologically-inspired cooperative multi-robot application. Specifically, a perimeter detection and tracking problem is discussed in detail. Finally, section 4 gives some concluding remarks and future directions.

2. The Multi-Robot Experimental Testbed

2.1 Platform Hardware

The platform consists of a R/C truck chassis from Tamiya Inc., odometer wheel sensors, a stereo vision system, an embedded computer (*e.g.,* , PC-104 or laptop), a suite of sensors, actuators, and wireless communication capabilities. The lower-level sensing, speed control, and actuator control are all networked using the Controller Area Network (CAN) system (Etschberger, 2001). This lower-level network is capable of controlling linear speed up to 2 m/s, angular speed, and managing the network. Having a velocity controlled platform is important, since many coordination approaches available in the literature (Das et al., 2002) have been developed considering kinematic rather than dynamic mobile robot models. Managing the network consists of detecting modules, controlling the state of a module, and maintaining network integrity. The vehicles (see Figure 1 left) are versatile enough to be used indoors, or outdoors in good weather.

Figure 1. (a) MARHES multi-vehicle experimental testbed, and (b) setup for perimeter detection.

2.2 System Architecture

The system architecture, shown in Figure 2, consists of low-level and high-level layers. The low-level layer is made up of sensors, electronics, and a PCMCIA card to interface the CAN bus with the high-level PC. The high-level layer is made up of PC's and the server, all of which are using Player/Gazebo (Gerkey et al., 2003).

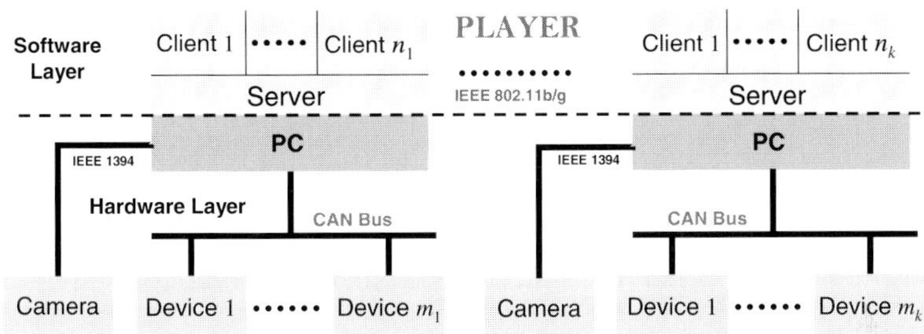

Figure 2. System architecture block diagram.

Player is a server for robotic communication, developed by USC, that is language independent and works under UNIX-based operating systems. It provides an interface to controllers and sensors/actuators over TCP/IP protocol. Each robot has a driver and is connected to a specified socket, which can *read/write* to all of the sensors and actuators on the robot. Users connect to the sockets through a client/robot interface and can send commands or receive information from each robot. Each robot can communicate with each other or a single common client and update their information continuously to all clients. Player is convenient in that it allows for virtual robots to be simulated by connecting the driver to Gazebo – a $3D$ open source environment simulator.

The CAN protocol is a message-based bus, therefore bases and workstation addresses do not need to be defined, only messages. These messages are recognized by message identifiers. The messages have to be unique within the whole network and define not only the content, but also its priority. Specifically, our lower-level CAN Bus system allows for sensors and actuators to be added and removed from the system with no reconfiguration required to the higher-level structure (*c.f.*, (Gomez-Ibanez et al., 2004)). This type of flexible capability is not common in commercially available mobile robotic platforms. The current configuration of the MARHES-TXT CAN system is shown in Figure 3.

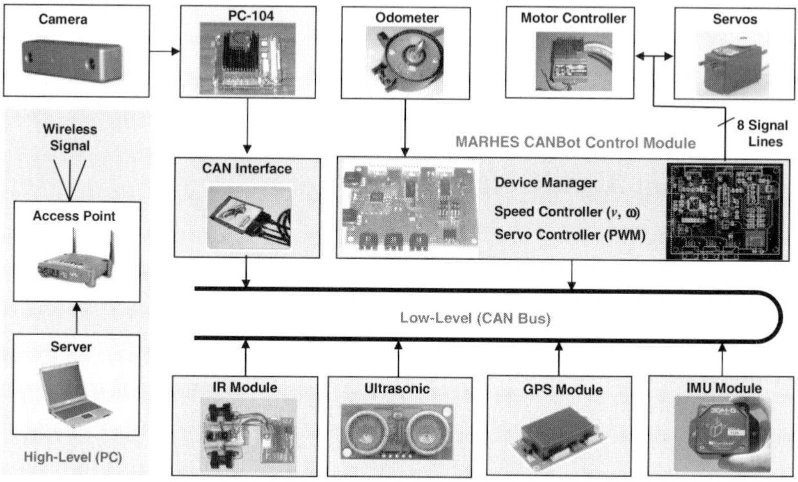

Figure 3. CAN Bus diagram.

The Device Manager Module (DMM) is the CAN component which manages the bus. After start-up, the DMM checks the bus continuously to see if a sensor has been connected/disconnected and that all components are working properly.

3. Biologically-Inspired Perimeter Detection and Tracking

Perimeter detection has a wide range of uses in several areas, including: (1) Military, *e.g.,* , locating minefields or surrounding a target, (2) Nuclear/Chemical Industries, *e.g.,* , tracking radiation/chemical spills, (3) Oceans, *e.g.,* , tracking oil spills, and (4) Space, *e.g.,* , planetary exploration. A perimeter is an area enclosing some type of agent. We consider two types of perimeters: (1) static and (2) dynamic. A static perimeter does not change over time, *e.g.,* , possibly a minefield. Dynamic perimeters are time-varying and expand/contract over time, *e.g.,* , a radiation leak.

A variety of perimeter detection and tracking approaches have been proposed in the literature. Bruemmer *et al.*, present an interesting approach in which a swarm is able to autonomously locate and surround a water spill using social potential fields (Bruemmer et al., 2002). Authors in (Feddema et al., 2002) show outdoor perimeter surveillance over a large area using a swarm that investigates alarms from intrusion detection sensors. A *snake* algorithm to locate and surround a perimeter represented by a concentration function is developed in (Marthaler and Bertozzi, 2004). Authors in (Savvides et al., 2004) use mobile sensing nodes to estimate dynamic boundaries.

In perimeter detection tasks a robotic swarm locates and surrounds an agent, while dynamically reconfiguring as additional robots locate the perimeter. Obviously, the robots must be equipped with sensors capable of detecting whatever agent they are trying to track. Agents could be airborne, ground-based, or underwater. See Fig. 4(a) for an example of a perimeter, an oil spill.

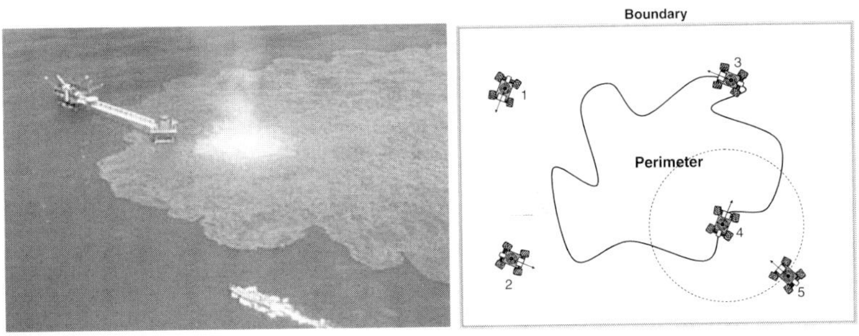

Figure 4. (a) Oil spill (Courtesy NOAA's Office of Response and Restoration), and (b) Boundary, perimeter example.

In this section, a decentralized, cooperative hybrid system is presented utilizing biologically-inspired emergent behavior. Each controller is composed of finite state machines, and it is assumed that the robots have a suite of sensors and can communicate only within a certain range. A relay communication scheme is used. Once a robot locates the perimeter, it broadcasts the location to any robots within range. As each robot receives the perimeter location, it too begins broadcasting, in effect, forming a relay. Other groups have used the terms *perimeter* and *boundary* interchangeably, but in this work, there is a distinct difference. The perimeter is the *chemical* agent being tracked, while the *boundary* is the limit of the exploration area. Refer to Fig. 4(b).

3.1 Cooperative Hybrid Controller

The theory of hierarchical hybrid systems (Alur et al., 2001, Fierro et al., 2002) offers a convenient framework to model the multi-robot system engaged in a perimeter detection and tracking task. At the higher layer of the hierarchy, the multi-robot sensor system is represented by a *robot-group* agent. Each robot agent is represented by a finite automaton consisting of three states as shown in Figure 5: (1) *searching*, (2) *moving*, and (2) *tracking*. The next layer of states are: (1) *collision avoidance* and (2) *boundary avoidance*. The lower layer is made up of elementary robot behaviors: (1) *speed up*, (2) *slow down*,

(3) *turn right*, (4) *turn left*, and (5) *go straight*. These actions change depending on which state the robot is in and if the sensors have detected the perimeter or neighboring robots.

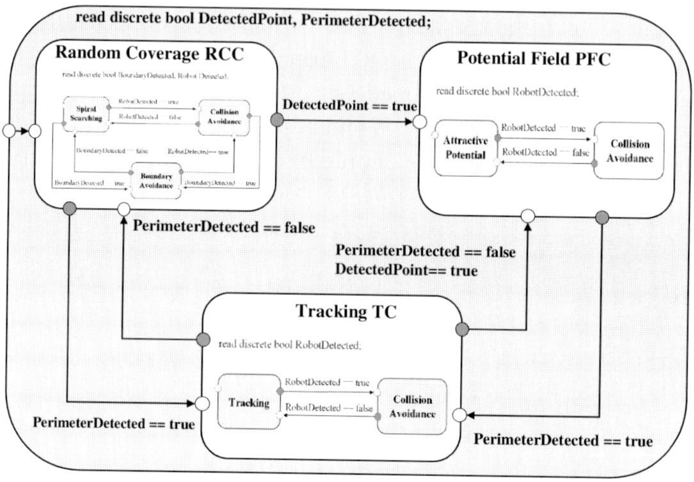

Figure 5. Mobile robot (sensor) agent.

The MARHES car-like platform is modeled with the unicycle model:

$$\dot{x}_i = v_i \cos\theta_i, \quad \dot{y}_i = v_i \sin\theta_i, \quad \dot{\theta}_i = \omega_i, \tag{1}$$

where x_i, y_i, θ_i, v_i, and ω_i are the x-position, y-position, orientation angle, linear velocity, and angular velocity of robot i, respectively. Note that v_i ranges from $0 \leq v_i \leq 2\ m/s$, while ω_i ranges from $-0.3 \leq \omega_i \leq 0.3\ rad/s$. These ranges come from extensive tests of our platform.

The controllers used are: (1) Random Coverage Controller (RCC), (2) Potential Field Controller (PFC), and (3) Tracking Controller (TC). The RCC uses a logarithmic spiral search pattern to look for the perimeter while avoiding collisions. The PFC allows the robots to quickly move to the perimeter if the perimeter has been detected. The TC allows the robots to track the perimeter and avoid collisions.

Random Coverage Controller. The goal of the Random Coverage Controller (RCC) is to efficiently cover as large an area as possible while searching for the perimeter and avoiding collisions. The RCC consists of three states: (1) Spiral Search, (2) Boundary Avoidance, and (3) Collision Avoidance. The spiral search is a random search for effectively covering the area. The boundary and collisions are avoided by adjusting the angular velocity.

The logarithmic spiral, seen in many instances in nature, is used for the search pattern. In (Hayes et al., 2002), a spiral search pattern such as that used by moths is utilized for searching an area. It has been shown that the spiral search is not optimal, but effective (Gage, 1993). Some examples are hawks approaching prey, insects moving towards a light source, sea shells, spider webs, and so forth. Specifically, the linear and angular velocity controllers are:

$$v_i(t) = v_s\left(1 - e^{-t}\right), \quad \omega_i(\theta_i) = ae^{b\theta_i}, \tag{2}$$

where $v_s = 1\ m/s$, a is a constant, and $b > 0$. If $a > 0$ (< 0), then the robots move counterclockwise (clockwise). If a robot gets close to the *boundary* (limit of the exploration area here), then it will turn sharply ($\omega_i = -\omega_{max}$, with $\omega_{max} = 0.3\ rad/s$) to avoid crossing the boundary. Once a certain distance away from the boundary is reached, then the controller will return to the spiral search.

To avoid collisions, the robots' orientations must be taken into account so the robots turn in the proper direction. If the robots are coming towards each other, then they both turn in the same direction to avoid colliding. If the *follower* robot is about to run into the *leader* robot, then the follower robot will turn, while the leader robot continues in whatever direction it was heading.

Potential Field Controller. Potential fields have been used by a number of groups for controlling a swarm. In (Tan and Xi, 2004), virtual potential fields and graph theory are used for area coverage. In (Parunak et al.,), a potential field method is used that is inspired by an algorithm observed in wolf packs.

The Potential Field Controller (PFC) uses an attractive potential which allows the robots to quickly move to the perimeter once it has been detected. The first robot to detect the perimeter *broadcasts* its location to the other robots. If a robot is within range, then the PFC is used to quickly move to the perimeter. Otherwise, the robot will continue to use the RCC unless it comes within range, at which point it will switch to the PFC. As a robot moves towards the goal, if it detects the perimeter before it reaches the goal, it will switch to the TC. The PFC has two states: (1) Attractive Potential and (2) Collision Avoidance.

The attractive potential, $\mathbf{P_a}(x_i, y_i)$, is

$$\mathbf{P_a}(x_i, y_i) = \frac{1}{2}\varepsilon[(x_i - x_g)^2 + (y_i - y_g)^2], \tag{3}$$

where (x_i, y_i) is the position of robot i, ε is a positive constant, and (x_g, y_g) is the position of the attractive point (goal). The attractive force, $\mathbf{F_a}(x_i, y_i)$, is derived below.

$$\mathbf{F_a}(x_i, y_i) = -\nabla \mathbf{P_a}(x_i, y_i) = \varepsilon \begin{bmatrix} x_g - x_i \\ y_g - y_i \end{bmatrix} = \begin{bmatrix} F_{a,x_i} \\ F_{a,y_i} \end{bmatrix} \tag{4}$$

Equation (4) is used to get the desired orientation angle, $\theta_{i,d}$, of robot i:

$$\theta_{i,d} = \arctan 2\left(\frac{F_{a,y_i}}{F_{a,x_i}}\right) \tag{5}$$

Depending on θ_i and $\theta_{i,d}$, the robot will turn the optimal direction to quickly line up with the goal using the following proportional angular velocity controller:

$$\omega_i = \begin{cases} 0 & \theta_i = \theta_{i,d} \\ \pm k(\theta_{i,d} - \theta_i) & \theta_i \neq \theta_{i,d}, \end{cases} \tag{6}$$

where $k = \frac{\omega_{max}}{2\pi}$ and $\omega_{max} = 0.3\ rad/s$.

Collisions are avoided in the same manner as in the random coverage controller.

Tracking Controller. The Tracking Controller changes ω and v in order to track the perimeter and avoid collisions, respectively. Cyclic behavior *emerges* as multiple robots track the perimeter. In (Marshall et al., 2004) , cyclic pursuit is presented in which each robot follows the next robot (1 follows 2, 2 follows 3, etc.). The robots move with constant speed and a proportional controller is used to handle orientation. The TC differs in that each robot's objective is to track the perimeter, while avoiding collisions. There are no restrictions on robot order, v is not constant, and a bang-bang controller is used to handle orientation.

The robots move slower in this state to accurately track the perimeter counterclockwise. The TC consists of two states: (1) Tracking, and (2) Cooperative Tracking. Tracking is accomplished by adjusting the angular velocity with a bang-bang controller. Cooperative tracking allows the swarm to distribute around the perimeter.

A robot only enters this state if it is the only robot that has detected the perimeter. The linear velocity is a constant 0.4 m/s. The angular velocity controller is:

$$\omega_i = \begin{cases} -\omega_t & \text{inside perimeter} \\ \omega_t & \text{outside perimeter}, \end{cases} \tag{7}$$

where $\omega_t = 0.1\ rad/s$. A look-ahead distance is defined to be the sensor point directly in front of the robots. If the sensor has detected the perimeter and the look-ahead distance is inside the perimeter, then the robot will turn right. Otherwise, the robot turns left. This zigzagging behavior is often seen in moths following a pheromone trail to its source.

Cooperative tracking occurs when two or more robots have sensed the perimeter. Equation (7) is still used, but now the linear velocity is adjusted to allow the swarm to distribute. The robots communicate their separation distances to each other. Each robot only reacts to its nearest neighbor on the perimeter. The swarm will attempt to uniformly distribute around the perimeter at a desired separation distance, d_{des}, using the following linear velocity proportional controller:

$$v_i = \begin{cases} 0 & |d_{ij} - d_{des}| < \varepsilon \\ k_p|d_{des} - d_{ij}| & \text{otherwise} \end{cases} \tag{8}$$

where d_{ij} is the distance from robot i to robot j, $\varepsilon = 0.01$, and $k_p = \frac{v_{max}}{|d_{des} - d_{ij,max}|} \simeq 0.06$.

When the swarm is uniformly distributed, it stops. This allows each robot to conserve its resources. If the perimeter continues to expand or robots are added or deleted, then the swarm will reconfigure until the perimeter is uniformly surrounded. Note that the choice of d_{des} is critical to the swarm's behavior. If d_{des} is too low, the swarm will not surround the perimeter. Subgroups may also be formed since each robot only reacts to its nearest neighbor. If d_{des} is too high, the swarm will surround the perimeter, but continue moving.

Collision avoidance is handled inherently in the controller. Since the robots can communicate, they should never get close enough to collide.

An analysis of the system is being done in order to calculate the optimal number of robots needed to uniformly distribute around a perimeter, assuming a static perimeter and a constant d_{des}.

Ideally, the swarm could dynamically change d_{des} depending on the perimeter. This should allow the swarm to uniformly distribute around most perimeters, static or dynamic. We are currently investigating ways to do this.

3.2 Results

In the simulation depicted in Figure 6, D_{com}, D_{sen}, and D_{sep} are the communication range, the sensor range, and the desired separation distance, respectively. Five robots are shown tracking a dynamic perimeter that is expanding at 0.0125 m/s.

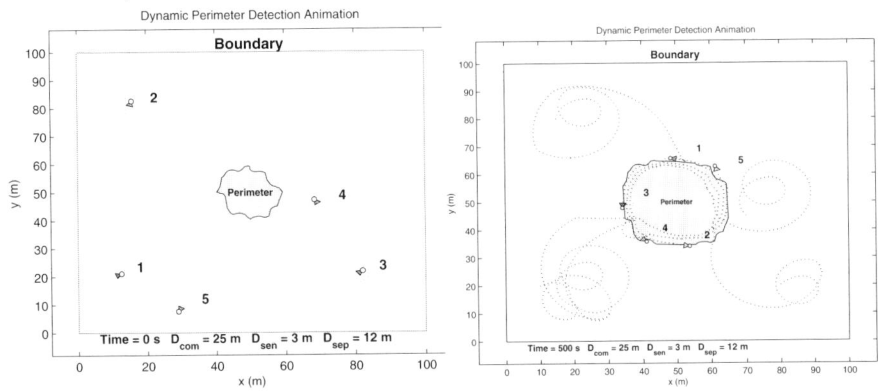

Figure 6. Five robots detecting and tracking a dynamic perimeter. (a) Initial and (b) final configurations.

All robots are initially searching. Robot 4 locates the perimeter first. Robot 3 is within range and receives the location. As robot 4 is tracking the perimeter, robot 2 comes within range, receives the location, and moves toward it avoiding collisions with robot 4. Robots 1 and 5 locate the perimeter on their own.

Subgroups are formed in this case because there are not enough robots to distribute around the perimeter. The swarm does not stop because the expanding perimeter is never uniformly surrounded.

A Gazebo model was developed to verify the simulation results from Matlab. Our platform, the Tamiya TXT-1, can be modeled in Gazebo using the ClodBuster model. Each robot is equipped with odometers, a pan-tilt-zoom camera, and a sonar array. Position and orientation are estimated using the odometers. The camera is used for tracking the perimeter. It is pointed down and to the left on each robot to allow the robots to track the perimeter at a small offset. This way, they never drive into the perimeter. The sonar array is used to avoid collisions.

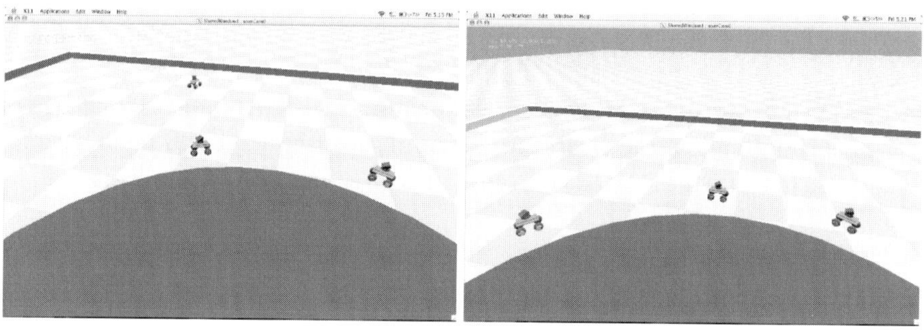

Figure 7. Perimeter detection and tracking using Gazebo.

A simulation is shown in Figure 7 in which robots search for, locate, and track the perimeter while avoiding collisions. The perimeter and boundary are represented by a red cylinder and a gray wall, respectively. Since Gazebo models the real world (sensors, robots, and the environment) fairly accurately, the code developed in Gazebo should be easily portable to our experimental testbed.

4. Summary

We describe a cost-effective experimental testbed for multi-vehicle coordination. The testbed can be used indoors under controlled lab environments or outdoors. Additionally, we design a decentralized cooperative hybrid system that allows a group of *nonholonomic* robots to search for, detect, and track a dynamic perimeter with only limited communication, while avoiding collisions and reconfiguring *on-the-fly* as additional robots locate the perimeter. Current work includes optimization based sensor placement for perimeter estimation, and implementation of the controllers presented herein on the MARHES experimental testbed.

Acknowledgments

This material is based upon work supported in part by the U. S. Army Research Laboratory and the U. S. Army Research Office under contract/grant number DAAD19-03-1-0142. The first author is supported in part by NSF Grants #0311460 and #0348637. The second author is supported in part by a NASA Oklahoma Space Grant Consortium Fellowship. We would like to thank Kenny Walling for his work on the mobile platform. Also, we thank Daniel Cruz, Pedro A. Diaz-Gomez, and Mark Woehrer for their work on Player/Gazebo.

References

Alur, R., Dang, T., Esposito, J., Fierro, R., Hur, Y., Ivancic, F., Kumar, V., Lee, I., Mishra, P., Pappas, G., and Sokolsky, O. (2001). Hierarchical hybrid modeling of embedded systems. In Henzinger, T. and Kirsch, C., editors, *EMSOFT 2001*, volume 2211 of *LNCS*, pages 14–31. Springer-Verlag, Berlin Heidelberg.

Bruemmer, D. J., Dudenhoeffer, D. D., McKay, M. D., and Anderson, M. O. (2002). A robotic swarm for spill finding and perimeter formation. In *Spectrum 2002*, Reno, Nevada USA.

Das, A. K., Fierro, R., Kumar, V., Ostrowski, J. P., Spletzer, J., and Taylor, C. J. (2002). A vision-based formation control framework. *IEEE Trans. on Robotics and Automation*, 18(5):813–825.

Etschberger, K. (2001). *Controller Area Network: Basics, Protocols, Chips and Applications*. IXXAT Press, Weingarten, Germany.

Feddema, J. T., Lewis, C., and Schoenwald, D. A. (2002). Decentralized control of cooperative robotic vehicles: Theory and application. *IEEE Trans. on Robotics and Automation*, 18(5):852–864.

Fierro, R., Das, A., Spletzer, J., Esposito, J., Kumar, V., Ostrowski, J. P., Pappas, G., Taylor, C. J., Hur, Y., Alur, R., Lee, I., Grudic, G., and Southall, J. (2002). A framework and architecture for multi-robot coordination. *Int. J. Robot. Research*, 21(10-11):977–995.

Gage, D. W. (1993). Randomized search strategies with imperfect sensing. In *Proceedings of SPIE Mobile Robots VIII*, volume 2058, pages 270–279, Boston, Massachusetts USA.

Gerkey, B. P., Vaughan, R. T., and Howard, A. (2003). The player/stage project: Tools for multi-robot and distributed sensor systems. In *Proc. IEEE/RSJ Int. Conf. on Advanced Robotics (ICAR)*, pages 317–323, Coimbra, Portugal.

Gomez-Ibanez, D., Stump, E., Grocholsky, B., Kumar, V., and Taylor, C. J. (2004). The robotics bus: A local communications bus for robots. In *Proc. of SPIE*, volume 5609. In press.

Hayes, A. T., Martinoli, A., and Goodman, R. M. (2002). Distributed odor source localization. *IEEE Sensors*, 2(3):260–271. Special Issue on Artificial Olfaction.

Jadbabaie, A., Lin, J., and Morse, A. S. (2003). Coordination of groups of mobile autonomous agents using nearest neighbor rules. *IEEE Trans. on Automatic Control*, 48(6):988–1001.

Marshall, J. A., Broucke, M. E., and Francis, B. A. (2004). Unicycles in cyclic pursuit. In *Proc. American Control Conference*, pages 5344–5349, Boston, Massachusetts USA.

Marthaler, D. and Bertozzi, A. L. (2004). Tracking environmental level sets with autonomous vehicles. In *Recent Developments in Cooperative Control and Optimization*. Kluwer Academic Publishers.

Parunak, H. V. D., Brueckner, S. A., and Odell, J. Swarming pattern detection in sensor and robot networks. Forthcoming at the 2004 American Nuclear Society (ANS) Topical Meeting on Robotics and Remote Systems.

Savvides, A., Fang, J., and Lymberopoulos, D. (2004). Using mobile sensing nodes for boundary estimation. In *Workshop on Applications of Mobile Embedded Systems*, Boston, Massachusetts.

Tan, J. and Xi, N. (2004). Peer-to-peer model for the area coverage and cooperative control of mobile sensor networks. In *SPIE Symposium on Defense and Security*, Orlando, Florida USA.

VI

HUMAN-ROBOT INTERACTION

TASK SWITCHING AND MULTI-ROBOT TEAMS

Michael A. Goodrich
Computer Science Department, Brigham Young University, Provo, UT, USA
mike@cs.byu.edu

Morgan Quigley
Computer Science Department, Brigham Young University, Provo, UT, USA.

Keryl Cosenzo
U.S. Army Research Laboratory, Aberdeen, MD, USA.

Abstract Determining whether it is possible for a single human to manage a team of multiple robots is an important question given current trends in robotics. Restricting attention to managing a team of multiple robots where a single human must be able to analyze video from each robot, we review how neglect time and interaction time of the interface-robot system provide a test for feasibility of a team. We then present a feasibility test that is applicable if the cost of switching attention between multiple robots or multiple tasks can become prohibitive. We then establish that switch costs can be high, and show that different tasks impose different switch costs.

Keywords: Switch costs, fan-out, human-robot interaction, multiple robot management

1. Introduction

Recently, there has been much discussion in the robotics community on creating robot systems that allow a single human to perform multiple tasks, especially managing multiple robots. The possibility for such one-to-many human-robot teams is caused by the ever-increasing autonomy of robots. As a robot becomes more autonomous, its human manager has more free time to do other tasks. What better way to use this free time than to have the human manage multiple robots or manage multiple tasks?

The potential impact of this line of reasoning includes some very desirable consequences, but there are some clear upper bounds on the number of robots and the number of tasks that a single human can manage. These upper bounds

L.E. Parker et al. (eds.),
Multi-Robot Systems. From Swarms to Intelligent Automata. Volume III, 185–195.

are created by how long a single robot can be neglected. Formally, *neglect time* is the expected amount of time that a robot can be ignored before its performance drops below a threshold.

During the time that a robot is being neglected, the human manager can conceivably be doing any other task. However, once the neglect time is exhausted, the human must interact with the robot again. The average amount of time required by the human to "retask" the robot once interaction begins is referred to as the interaction time. Formally, *interaction time* is the expected amount of time that a human must interact with a robot to bring it to peak performance.

In a problem with multiple robots, neglect time and interaction time dictate the maximum number of robots that a single human can manage. The upper bound on the number of robots can easily be computed when all robots are homogeneous and independent. The idea of determining how many independent homogenous robots can be managed by a single human is captured by the notion of fan-out (Olsen and Goodrich, 2003). Roughly speaking, fan-out is one plus the ratio of neglect time to interaction time. The ratio represents the number of other robots that the human can manage during the neglect time interval, and the "plus one" represents the original robot. Thus,

$$FanOut = \frac{NT}{IT} + 1$$

where NT and IT represent neglect time and interaction time, respectively.

This idea can be extended to teams of heterogeneous robots performing independent tasks. When a team is made up of heterogeneous robots, then each robot has its autonomy level and interface. This, in turn, implies that each robot has a given neglect time and interaction time. Let $N_i = (NT_i, IT_i)$ denote the neglect and interaction time of robot i. A team of M robots consists of the set $\mathcal{T} = \{N_i : i = 1 \ldots M\}$.

To determine whether a human can manage a team of robots $\mathcal{T}$, we can use the neglect times and interaction times to determine if a team is infeasible.

$$\mathcal{T} \text{ is } \begin{cases} \text{feasible} & \text{if } \forall i \;\; NT_i \geq \sum_{j \neq i} IT_j \\ \text{infeasible} & \text{otherwise} \end{cases} . \tag{1}$$

The key idea is to find out whether the neglect time for each robot is sufficiently long to allow the human to interact with every other robot in the team. If not, then the team is not feasible. If so, then there is sufficient time to support the team, though the team performance may be suboptimal, meaning that a different team configuration could produce higher expected performance.

Fan-out and the feasibility equation are upper bounds on the number of independent robots that can be managed by a single human. The purpose of this paper is demonstrate that the amount of time required to switch between robots can be substantial, and can dramatically decrease this upper bound.

2. Related Literature

Approaches to measuring switch costs are usually loosely based on fundamental models of cognitive information processing (Meiran et al., 2002, Leviere and Lee, 2002). These models suggest that procedural memory elements, encoded as modules in long-term memory and sometimes referred to as mental models, dictate how particular stimuli are interpreted and acted upon by a human. When the nature of the task changes, a required switch in mental models is required and this switch comes at a cost even if the stimuli does not change. Reasons for this cost include the need to prepare for the new task and inhibit the old task.

The experimental methodology typically adopted in the cognitive science literature has been to use the same set of stimuli but switch the task to be done on the stimuli(Cepeda et al., 2001, Koch, 2003). For example, the digits "1 1 1", "1", "3 3 3", and "3" can be randomly presented to a subject. One task requires the subject to name the digit (one, one, three, and three, respectively), and the other task requires the subject to count the number of digits depicted (three, one, three, and one, respectively). Switch cost is given by the extra time required when a trial requires a change from one task to another as compared to a trial when the task does not change.

This approach has limited application to the human-robot interaction domain for two reasons. First, the absolute values of the switch costs are very low; they are on the order of fifty milliseconds. Second, human-robot interaction domains are not simple stimuli-response tasks, but rather require the use of short term memory and the possible recruitment of multiple mental models to solve a problem. As a result, new experimental methodologies must be created to measure switch costs in human-robot interaction domains.

Altmann and Trafton have proposed one technique for measuring switch costs in more complex domains (Altmann and Trafton, 2004). Their approach, which has been applied to problems that impose a heavy burden on working memory and cognitive processing, is to measure the amount of time between when a switch is made to a new task and the first action is taken by the human on this new task. In their research, they measure the time between when the working environment causes a switch to a new task and when the human takes their first action on the new task. They have found that switch costs in complicated multi-tasking environments can be on the order of two to four seconds; this amount can impose serious limitations on the number of robots that a single human can manage.

It is important to note that Altmann and Trafton's research has included a study of signaling the human of an impending interrupt. They have found that signaling reduces switch costs (on the order of two seconds) because it allows people to prepare to resume the primary task when the interruption is completed. Their experiments suggest that people's preparation includes both retrospective and prospective memory components.

Unfortunately, the experiment approach used by Altmann and Trafton does not go far enough into naturalistic human-robot interaction domains to suit our needs. The primary limitation is the primary-secondary task nature of their experiments. Multi-robot control will not always have a primary robot with a set of secondary robots. A second limitation is the use of "first action after task resumption" as a metric for switch costs. In the multi-robot domain, a person may not take any action when a task is resumed either because the robot is still performing satisfactorily or because the human does not have enough situation awareness to properly determine that an action is required. Despite its limitations, we adopt the primary-secondary task approach to establish that switch costs can be high. However, we use a change detection approach for measuring recovery time.

3. Switch Costs

The preceding discussion has assumed that interaction time captures all interaction costs. Unfortunately, there is a cost associated with switching between multiple activities. This cost has been studied extensively under the name of "task switching" in the cognitive science literature, but has received considerably less attention in the human-robot interaction literature.

The problem with the preceding discussion of fan-out and feasibility is that it assumes no interference effects between tasks. Olsen noted this limitation in the definition of interaction time, and used the more general notion of interaction effort to include the actual time spent interacting with the robot as well as the time required to switch between tasks (Olsen, Jr. and Wood, 2004, Olsen, Jr. et al., 2004). Unfortunately, this work did not research the costs of task switching and does not, therefore, allow us to make predictions about the feasibility of a human-robot system or diagnose the problems with a given system.

Switch costs are important in domains where a human must manage multiple robots because managing multiple robots entails the need to switch between these robots. If the costs due to switching are significant, then the number of robots that can be managed dramatically decreases. As autonomy increases and interfaces become better, switch costs may become the bottleneck which limits the number of robots that a single human can manage. For example, suppose that a human is managing a set of robots that can be neglected for no more than 20 seconds. If the human is asked to manage a team of robots where each robot requires no more than five seconds of interaction time per interaction event, then the human can manage no more than five robots. If, however, switching between robots comes at a cost of, say, three extra seconds, then rotating between the five robots requires 15 seconds of switch cost which consumes 75% of the total neglect time without actually performing any useful interaction. This switch cost makes interaction effort jump from five seconds to eight seconds, and means that the human can manage at most three robots.

Formally, we denote the cost to switch between task i and task j as $\mathrm{SC}(i, j)$ where this cost has units of time; large times mean high costs. When a human begins to neglect task k, the feasibility constraint in Equation (1) demands that all the interaction times of all other tasks $j \neq k$ can be accomplished during the neglect time for task k. Since the experiment results strongly indicate that the switch cost can vary substantially depending on the type of secondary task, it is necessary to address how feasibility is affected by switch costs.

To this end, it is necessary to distinguish between what constitutes an *interaction time* and what constitutes a *switch cost*. The precise differentiation between these terms could be a topic of heated debate, but for our purposes we will use operational definitions of the two terms that are compatible with the experiment. The term *switch cost* denotes the amount of time between when one task ends and the operator demonstrates an ability to detect changes in the environment. This relies on the association between the *change detection* problem and situation awareness which we will discuss shortly. A good description of change detection can be found in Rensink (Rensink, 2002). "Change detection is the apprehension of change in the world around us. The ability to detect change is important in much of our everyday lifefor example, noticing a person entering the room, coping with traffic, or watching a kitten as it runs under a table." The term *interaction time* denotes the amount of time that is required for the operator to bring the robot to peak performance after a level of situation awareness is obtained that is high enough to detect changes in the environment.

The experiment results presented below indicate the the time to detect changes is sensitive to the type of tasks involved. Therefore, the feasibility equation must be modified to account for the effects of changes. The problem with doing so is that the total switch costs depends on the order in which the tasks are performed. Addressing this issue completely is an area of future work. For now, we adopt the most constraining definition and consider worst case switch costs.

Let $\mathcal{S}(i) = \{1, 2, \ldots, i-1, i+1, \ldots, n\}$ denote the set of all tasks different from task i. Consider the set of permutations over this set,

$$\mathcal{P}(i) = \{\text{permutations over } \mathcal{S}(i)\},$$

and let π denote a particular permutation within this set; $\pi(1)$ denotes the first element in the permutation, and so on. Let SC_i^* denote the largest possible cumulative switch cost for a given set of tasks in $\mathcal{S}(i)$, that is

$$\mathrm{SC}_i^* = \max_{\pi \in \mathcal{P}(i)} \mathrm{SC}(i, \pi(1)) + \sum_{k=1}^{n-2} \mathrm{SC}(\pi(k), \pi(k+1)) + \mathrm{SC}(\pi(n-1), i).$$

Note that we have included both the cost to switch from primary task i to the first permutation task in the permutation and the cost to switch from the last task in the permutation to the primary task. This is necessary because beginning the set of secondary tasks comes at a cost, and resuming the primary task also comes at a cost.

Feasibility of the team is then given by

$$\mathcal{T} \text{ is } \left\{ \text{ feasible} \quad \text{if } \forall i \;\; NT_i \geq \Sigma_{j \neq i} IT_j + \mathrm{SC}_i^* \right. . \tag{2}$$

If, for all tasks, the neglect time exceeds the sum of the interaction times plus the worst case switch costs, then the team is feasible.

In the next section, we describe an experiment that demonstrates that switch costs can be high enough (on the order of 5 to 10 seconds) to merit their consideration in determining team feasibility. We also show that type of switch is also an important consideration because various types of secondary tasks have substantially different switch costs.

4. The Experiment

We adopt the primary task/secondary task formulation in the experiment. The primary task was to control a simulated ground robot using a conventional display. This display included a video feed from the robot and a plan-view map of the environment. The environment consisted of treeless grass with multiple folds and hills.

Throughout the environment, there were ten geometric shapes randomly dispersed. Subjects used a gamepad to teleoperate the robot to within a meter of the geometric shapes. They then cycled through a set of geometric shapes (sphere, cube, or tetrahedron) by repeatedly clicking on one of the gamepad's buttons. The selected categorization was shown on the map view by placing a corresponding symbol on the map.

We adopted a change detection approach to indirectly measuring situation awareness. On approximately 50% of the trials (so that people will not be cued of a change), one of the geometric shapes changes or disappears from the camera view while the subject is performing the secondary task. The subject will be informed that this may occur in some trials, and will be asked to "alert their boss" that something has changed as soon as they detect a change. Alerting consisted of clicking one of two buttons to indicate the presence or absence of a change.

Figure 1. The before and after shots from the task switching experiment.

The experiment setup is illustrated in Figure 1 using screen shots from the experiment. The top figure shows the camera view (left) and map view (right) along with the geometric shapes (camera view) and their correct categorization (map view). In the top figure, a sphere is prominently displayed in the camera view. The corresponding sphere is circled in the map view to highlight its location in the map. (The circle is not shown to the subject, but is included in the figure to help highlight an important aspect of the experiment.)

The bottom figure shows the same world after the subject returns from the secondary task. Note how the sphere has disappeared from the camera view. Note further that the map view retains the information about the sphere. Thus, although the camera view has changed during the secondary task, the map view has the same information before and after.

The subject's task is to indicate that something changed while they were performing the secondary task. If a person has a good situation awareness after the secondary task, then they should be able to quickly consult the camera view to detect a change. If the situation awareness is poor, then they will need to compare the camera and map views to determine if something has changed. This forces the subject to "reconstruct" situation awareness and takes longer to perform. Secondary tasks that interfere with situation awareness should require the subject to take a longer time to recover.

Measuring the reaction time to detect this change after the task is resumed is an estimate of situation awareness. The time required to detect a change is an estimate of the time to achieve Endsley's "level 1" situation awareness (Endsley, 1997). Differences in times caused by various secondary tasks indicate different switch costs.

When subjects indicate that a change has occurred, we inform them whether they were correct. If they correctly identified a change, we require the subject to generate a report of what things have changed. The time to generate this report and the accuracy of the report will form a second measure of switch costs that we will analyze in future work. The nature of the report will be an updated categorization of all geometric shapes in the robot's camera field of view. This report will be made by removing missing shapes and recategorizing shapes that have changed. This report will be made by requiring the subject to (a) click on the shape in the map view that has disappeared if required, and (b) drive to the shapes that have changed or been added and (re)categorize them.

We experimented with four different types of secondary tasks.

- Blank screen: the screen goes blank for a preselected period of time.
- Tone counting: subjects are given a target tone and asked to count the number of times this target tone occurs in a two tone sequence. At the end of the sequence, subjects report the number of tones by clicking on the appropriate number in the display.
- Vehicle counting: subjects are asked to watch a video from a camera mounted on a real or simulated UAV, and to count the number of unique cars observed from the UAV. At the end of the sequence, subjects report the number of vehicles by clicking on the appropriate number in the display.
- Spatial reasoning (tetris): subjects are asked to play a game of tetris for a preselected period of time.

The blank screen serves as the baseline, both tone counting and vehicle counting place some burden on attention and working memory, and both vehicle counting and spatial reasoning place some burden on visual short term memory. Secondary tasks last between 10 seconds and 40 seconds. Tasks are presented in a balanced randomized schedule, and changes are generated randomly.

5. Results

For this paper, we estimate switch costs by measuring the amount of time between when the secondary task ends and when the subject pushes the button indicating that a change has occurred. Results are presented only for those conditions where a change occurred and the subject correctly identified the change. Future work should carefully address error rates as well as the sensitivity of these error rates and switch costs to the frequency with which changes occur.

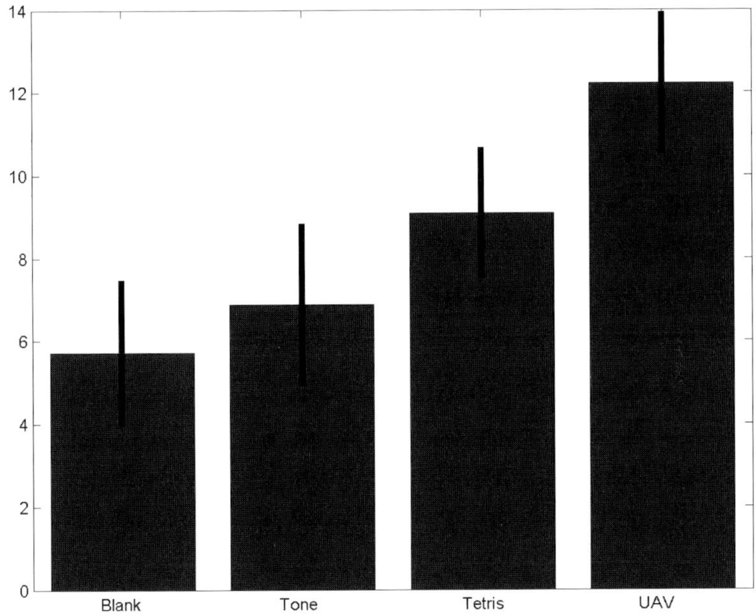

Figure 2. Average switch costs as a function of task with 20% confidence intervals.

The results of the experiment are shown in Figure 2 which displays the average switch costs across five subjects and seven one hour experiment sessions. Also shown are the 20% confidence intervals.

Two important things are worth noting. First, note that the average values for the switch costs range from over five seconds to just over twelve seconds. This is important because it indicates that switch costs can be very large. This indicates that an evaluation of the feasibility of a multi-robot team with independent robots should include an anaylsis of switch costs.

Second, note that the switch costs associated with the UAV are twice as large as the switch costs associated with the tone counting and blank screen. This indicates that there is a potentially large difference in the switch costs between various types of tasks. In fact, a two-sided t-test indicates that the different tasks all have statistically significant differences at the 20% level (or below) except for the difference between tone counting and tetris which appears to not be statistically significant. This data must be taken with a grain of salt because we only have five subjects (and seven total one-hour experiments) and only between 22 and 33 correct detections of changes (depending on the secondary task). However, this analysis combined with the magnitude of the effect strongly suggests that the different secondary tasks have a substantial influence on the switch costs.

Future work should carefully analyze why the different secondary tasks have such different switch costs. It is apparent that the differences cannot simply be attributed to counting since both the UAV and tone-counting tasks require the subjects to count, but these two tasks have different switch costs. It is also apparent that the differences cannot simply be attributed to visual overload since both the UAV and tetris are visual tasks, but these two tasks have different switch costs. Although there is not enough data to conclude that the duration of the secondary tasks is unimportant, there does not appear to be a big difference in the switch costs between tasks lasting fifteen seconds and tasks lasting thirty seconds.

We hypothesize that the differences in switch costs are attributable to load on working memory plus some component of spatial reasoning. This suggests that the feasibility of a team where a single human must analyze the video from multiple independent robots should be carefully studied.

It is important to note that at the end of the experiment, we asked subjects to report an estimate of the relative subjective workload of the various tasks. We did this by asking them if one secondary task was easier to recover from than another. All subjects reported that all four tasks were equal in this regard. We hypothesize that this estimate results from the fact that confirming when no change was made requires an almost exhaustive search through the map-view. Importantly, the subjective evaluations of workload did not correspond to the actual performance on the tasks.

6. Conclusions and Future Work

The experiment suggests that switch costs can have a substantial influence on the total cost of managing multiple tasks, and that the switch costs depend to some extent on the nature of the secondary task. We can include the effects of these switch costs by estimating the worst case switch cost for multiple secondary tasks. This worst case can then be used to identify obviously infeasible teams. Future work should explore the efficient computation of these switch costs, and the difference between the worst case feasibility and actual feasibility. Future work should also explore how intelligent interfaces and robot autonomy could be designed to minimize switch costs and support recovery from interruptions.

References

Altmann, E. M. and Trafton, J. G. (2004). Task interruption: Resumption lag and the role of cues. In *Proceedings of the 26th annual conference of the Cognitive Science Society*.

Cepeda, N. J., Kramer, A. F., and de Sather, J. C. M. G. (2001). Changes in executive control across the life span: Examination of task-switching performance. *Developmental Psychology*, 37(5):715–730.

Endsley, M. R. (1997). The role of situation awareness in naturalistic decision making. In Zsambok, C. E. and Klein, G., editors, *Naturalistic Decision Making*, chapter 26, pages 269–283. Lawrence Erlbaum Associates, Hillsdale, N.J.

Koch, I. (2003). The role of external cues for endogenous advance reconfiguration in task switching. *Psychonomic Bulleting and Review*, 10:488–492.

Leviere, C. and Lee, F. J. (2002). Intention superiority effect: A context-switching account. *Cognitive Systems Research*, 3:57–65.

Meiran, N., Hommel, B., Bibi, U., and Lev, I. (2002). Consciousness and control in task switching. *Consciousness and Cognition*, 11:10–33.

Olsen, D. R. and Goodrich, M. A. (2003). Metrics for evaluating human-robot interactions. In *Proceedings of PERMIS 2003*.

Olsen, Jr., D. R. and Wood, S. B. (2004). Fan-out: measuring human control of multiple robots. In *Proceedings of the 2004 conference on Human factors in computing systems*, pages 231–238. ACM Press.

Olsen, Jr., D. R., Wood, S. B., and Turner, J. (2004). Metrics for human driving of multiple robots. In *Proceedings of the 2004 IEEE Intl. Conf. on Robotics and Automation*, volume 3, pages 2315–2320.

Rensink, R. (2002). Change detection. *Annu. Rev. Psychol.*, 53:245–277.

USER MODELLING FOR PRINCIPLED SLIDING AUTONOMY IN HUMAN-ROBOT TEAMS

Brennan Sellner, Reid Simmons, Sanjiv Singh
Robotics Institute
Carnegie Mellon University
5000 Forbes Avenue, Pittsburgh, PA 15213
*United States of America**
bsellner@andrew.cmu.edu, reids@cs.cmu.edu, ssingh@ri.cmu.edu

Abstract

The complexity of heterogeneous robotic teams and the domains in which they are deployed is fast outstripping the ability of autonomous control software to handle the myriad failure modes inherent in such systems. As a result, remote human operators are being brought into the teams as equal members via sliding autonomy to increase the robustness and effectiveness of such teams. A principled approach to deciding when to request help from the human will benefit such systems by allowing them to efficiently make use of the human partner. We have developed a cost-benefit analysis framework and models of both autonomous system and user in order to enable such principled decisions. In addition, we have conducted user experiments to determine the proper form for the learning curve component of the human's model. The resulting automated analysis is able to predict the performance of both the autonomous system and the human in order to assign responsibility for tasks to one or the other.

Keywords: Mixed Initiative, User Modelling, Sliding Autonomy, Multiagent, Cooperation

1. Introduction

As complex robotic systems are deployed into ever more intricate and real-world domains, the demand for system abilities is growing quickly. Since many tasks cannot be easily accomplished by a single machine, much research has turned towards utilizing heterogeneous robotic teams. While this approach multiplies the theoretical capabilities of the deployed hardware, the actual abil-

*This work is partially supported by NASA grant NNA04CK90A.

L.E. Parker et al. (eds.),
Multi-Robot Systems. From Swarms to Intelligent Automata. Volume III, 197–208.

ities of a team often are constrained by its control software. In complex tasks, such as those required by automated assembly domains, it is nearly impossible for the system designer to anticipate every possible system failure and provide a method for recovery. While automated systems excel at rapid repetition of precise tasks, they are weak when dealing with such unexpected failures. As a result, research is now moving towards including a human in such teams, leveraging the accuracy and strength of robotic teams and the flexibility of the human mind to create a whole greater than the sum of its parts.

A great difficulty in creating these sliding autonomy systems is enabling smooth and efficient transitions between modes of autonomy - ideally both the human and the autonomous system should be able to initiate such transitions as they see fit. If the system is to do so, it needs some method for making decisions about when and how to involve the human in its task. The approach we have taken is to form models of the capabilities of both the autonomous system and the human, in order to provide a principled basis for the system to perform cost-benefit analysis. The autonomous system does not learn to improve its task performance, resulting in a model based on a static distribution derived from observed data. The human model is similar, but incorporates an explicit model of the human's learning curve, allowing the system to predict future performance of a human still learning a particular task. We have experimentally determined that a logarithmic function provides a good fit to our subjects' actual learning curves, with the model providing useful predictions during the learning period. Coupled with a cost-benefit analysis framework, these models allow the system to estimate the overall expected cost of transferring control to the human at various points during the task, enabling it to proactively involve the human when the human will provide the team with significant assistance.

2. Related Work

Our Syndicate architecture (Sellner et al., 2005) (Simmons et al., 2002) (Goldberg et al., 2003) provides a flexible, tiered, multi-agent architecture which we have extended to support sliding autonomy. Syndicate differs from most other multi-robot architectures by allowing close coordination without the need for a central planner. Our user modelling implementation continues this decentralization by allowing each agent to individually form models for itself and the human performing tasks using that agent's hardware.

A number of other sliding autonomy systems exist, of greater or lesser similarity to our work. (Fong et al., 2003) enable the robot to ask the operator for help with localization and to clarify sensor readings, while the operator can query the robot for information. This framework uses the human as an information source, rather than a true partner, and assumes the robot's control software is capable of performing all tasks when provided with complete state information. Our approach allows the operator to be a partner in the completion of the scenario, rather than simply a source of information. An architecture for sliding autonomy as applied to a daily scheduler has been proposed by (Scerri and Pynadath, 2002). The autonomous system is responsible for resolving timing conflicts among team members, who are able to adjust the system's autonomy by indicating intent or willingness to perform tasks. Using similar hardware to ours, (Kortenkamp et al., 1999) have developed and tested a software architecture that allows for sliding autonomous control of a robotic manipulator. While these projects all involve the human in the task, they do not explicitly reason about when to request help.

Similar to our modelling approach, (Fleming and Cohen, 2001) perform cost-benefit calculations to determine whether an agent should ask the user for information that may allow it to generate better plans. Although the basic cost-benefit concept is the same, our user models differ significantly. They represent the user by a series of ad-hoc probabilities (such as the probability that the user will have the requisite knowledge to answer a question), expected utilities, and costs. Their work does not consider the problem of user model acquisition, which is clearly far from trivial. In addition, their agent queries the user only when it believes that it needs help and that the user can provide the requisite information. There is no concept of ceding control to the user merely because the user is better at some element of the task; instead, the user is again treated as an information source, rather than as a partner.

Our sliding autonomy implementation allows any component of our multi-layered system to be switched between autonomous and manual (tele-operated) modes. The fine granularity of control over the team's autonomy level afforded by this approach allows many combinations of human intuition and robotic calculation, rather than limiting the human to the role of oracle. This switching may be performed in three ways: (1) pre-scripted, such as tasks which the autonomous system had not been programmed to perform and must be completed by the operator, (2) human-initiated changes in autonomy resulting from the operator deciding he wants to take control, and (3) system-initiated autonomy changes, which occur when the system's analysis indicates the benefits of requesting help would outweigh the costs. This allows a synergy of human flexibility and robotic accuracy which yields a team with greater efficiency and reliability than either a purely autonomous or purely tele-operated approach. See (Brookshire et al., 2004) for a discussion of our implementation of sliding autonomy and related experimental results.

3. The Task

For our work on architectures, sliding autonomy, and user modelling, we developed several assembly scenarios that require close coordination between disparate agents. The scenario discussed here requires the team to assemble a square from four beams and four planarly compliant nodes (Figure 1d). The nodes are free to move about in the plane of the workspace, in a weak parallel to orbital assembly. When a beam is inserted into a node, enough force is required to cause an unconstrained node to roll away, rather than the beam's latches engaging the node. In order to provide a countervailing force, the team must brace each node while inserting every beam. To further complicate matters, neither of our manipulator agents possess any extrinsic sensors.

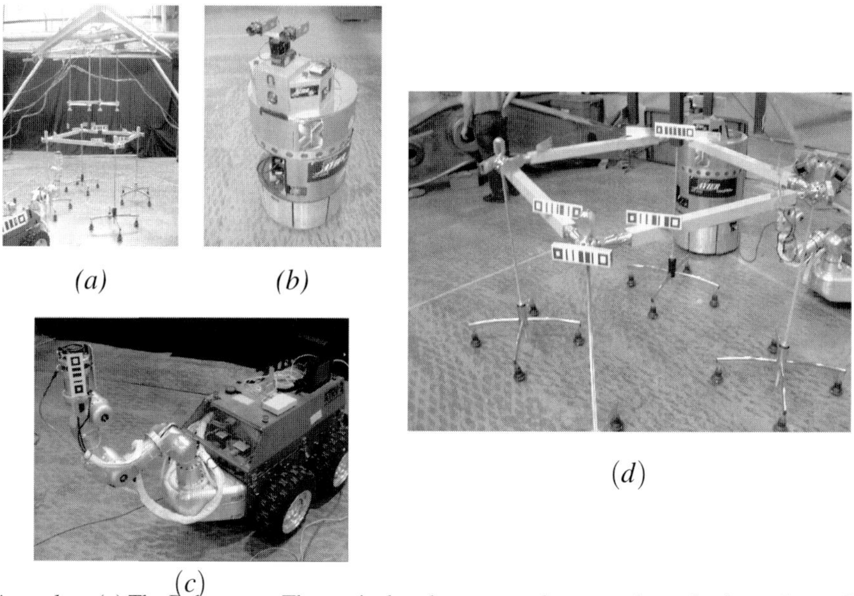

Figure 1. **(a)** The Robocrane. The vertical sockets are used to grasp the nodes from above. **(b)** Xavier, the roving eye of our work crew. **(c)** The mobile manipulator is composed of Bullwinkle (the differential-drive base) and Whiplash (the 5 degree-of-freedom anthropomorphic arm). **(d)** A closeup of the completed structure.

Thus, the scenario can be naturally split into three duties: docking, bracing, and sensing. Our mobile manipulator (Figure 1c) is responsible for docking the beams to the nodes with its 5-DOF anthropomorphic arm. The crane (Figure 1a) handles bracing, while the roving eye (Figure 1b) is responsible for providing information to the other agents about the relative positions of objects in the workspace. Each of these three agents independently performs cost-benefit analysis to determine whether it should ask for human assistance during the scenario.

The scenario consists of four repetitions of the following:

a Grasp beam with mobile manipulator's arm.

b Acquire beam and node with roving eye's sensors.

c Position and brace first node with crane.

d Insert one end of beam into first node by visually servoing mobile manipulator's arm.

e Reposition roving eye to second end of beam and acquire beam and second node.

f Release crane's grasp on first node, grasp second node, position node, and brace it.

g Insert second end of the beam into node.

h Release mobile manipulator's grasp on beam.

i Move mobile manipulator to the square's next side.

The team is *able* to accomplish the entire task autonomously, except for step 1, in which a human places the beam in the mobile manipulator's grasp. However, the human also may become involved in any of the other steps. If during step 3 or 6 the crane becomes stuck, the operator will need to intervene, since the current system cannot detect this failure. In step 2 or 5, if the roving eye is in a position such than an object of interest is obscured, the autonomous system will be unable to acquire, and will request assistance from the user. Docking one end of a beam to a node (steps 4 and 7) is a difficult task; the system will often fail one or more times before succeeding or dropping the beam. This is another opportunity to involve the operator, since an initial failure of the autonomous system is a good predictor of future failure; this occurence often results in a request for help after one or two failures.

Although the scenario can be accomplished autonomously, there are many opportunities for the system to request help from the human operator to increase its robustness and efficiency.

4. Using the User

The original sliding autonomy system we created was effective, but somewhat lacking in initiative. The system requested help only when told to do so ahead of time or when the team detected a failure from which it could not recover. This is clearly suboptimal: in the ideal case the autonomous system should request help not only when it *needs* assistance, but also when assistance would be beneficial to the reliable and efficient completion of the scenario. For instance, if the system has a failure recovery procedure for a particular error, but the procedure proves ineffective, it could ask the user for help after determining that further attempts are likely to be useless, rather than repeatedly attempting to blindly apply its recovery procedure. The node-beam docking action (steps 4 and 7 above) is an excellent example of this. In addition, there are occasionally tasks which the human is often more efficient at performing via tele-operation than the system, due to her superior ability to make inferences from noisy observations. Such tasks within our scenario include manuvering in cluttered environments and visually searching for partially obscured objects.

If the system is to further involve the human in the scenario, it must have some method of reasoning about when to do so. The approach that we have taken is to perform cost-benefit analysis at various decision points during the scenario, using empirically derived models of the individual users and the autonomous system to inform the analysis. By maintaining such individual models, the system's requests for help may depend on the abilities and state of the individual operator, yielding a team that adapts not only to the current state of the environment but also to the current state of its members. Such a principled approach allows the autonomous system to leverage the skills of the human operator to increase both the team's robustness and its efficiency.

4.1 Cost-Benefit Analysis

Cost-Benefit Analysis simply consists of estimating the costs and benefits associated with various courses of action in order to choose the most beneficial action to perform. In our domain, such decisions are binary: the system must decide whether to request operator assistance for a particular task. Obviously, the option with the greatest $benefit - cost$ value will be chosen. Given methods for estimating the relevant variables, this provides a framework for making principled decisions, rather than implementing arbitrary policies. Within our robotic assembly scenarios, one form of this equation is:

$$\begin{aligned} cost: &\quad price(h)E(t_h) + price(r_t)E(t_h) + price(rep)P(fcat_h) \\ benefit: &\quad price(r_a)E(t_r) + price(rep)P(fcat_r) \end{aligned} \tag{1}$$

where:

$E(t_h)$: Expected time for human to complete task
$E(t_r)$: Expected time for autonomous system to complete task
$P(fcat_h)$: Probability of catastrophic failure while under human control
$P(fcat_r)$: Probability of catastrophic failure while under autonomous control
$price(rep)$: Average monetary cost of repairing a catastrophic failure
$price(h)$: Monetary cost of operator per unit time
$price(r_t)$: Monetary operating cost of system per unit time while tele-operated
$price(r_a)$: Monetary operating cost of system per unit time while under autonomous control

The costs are those incurred during the human's teleoperation of the system, while the benefits consist of the cost savings associated with not running the system under autonomous control. In a real-world application, the price functions would be informed by factors such as the amortized cost of the hardware, upkeep, the salary of the human operator, and what other duties he is responsible for (since assisting the system will monopolize his time). These functions act as gains, with the relative values of $price(h)$, $price(r_t)$, and $price(r_a)$ encouraging or dissuading the system from asking for help, and $price(rep)$ adjusting how averse the system is to risk . The probability of catastrophic failure is estimated from experimental data. Note that catastrophic failure is distinct from the failure to perform a task in that the former results in damage to the robots which must be repaired while the latter merely results in the non-accomplishment of a particular task.

The most difficult element of these equations to estimate is the expected time to complete a given task for both the autonomous system and the human ($E(t_r)$ and $E(t_h)$, respectively), especially if the operator is a novice. We have built a user model to estimate these expected times based on previous experience, as well as a number of other factors.

4.2 User Model

A user model can consist of any collection of rules or set of assumptions that predicts the value of interest or otherwise allows the system to decide when to involve the user. In fact, our initial sliding autonomy system incorporated an extremely basic user model by requesting help only when an unrecoverable failure occurred. This simple approach allowed the user to slightly increase the system's robustness, but not its efficiency. A more refined model could include fixed thresholds for when help should be requested. Such a model could assert that if the system has failed to recover from an error twice it should ask for help. Again, this allows the user to contribute to the system's robustness, but the human is likely not being utilized in an efficient manner. In order to create an efficient overall system and take into account external constraints on the human, a much more detailed and data-driven model is required.

The Ideal Model. We have developed predictive models for both the autonomous system and the human operator; we address the system's model first. Since our current autonomous system does not learn, we may treat each attempt at a task as a sample from a static distribution. The set of all observed performances is in all likelihood multimodal. However, by segmenting the observed attempts based on the outcome of each attempt and the number of times the system had previously failed to perform the task during the current trial, we may easily form a set of unimodal (or nearly unimodal) distributions. We may then estimate $E(t_r)$ directly from these distributions:

$$E(t_r|F_r=i) = \begin{array}{l} P(S_r|F_r=i)E(t_r|S_r,F_r=i) \\ +P(\neg S_r|F_r=i)\left(\begin{array}{l} E(t_r|\neg S_r,F_r=i) \\ +E(t_r|F_r=i+1) \end{array}\right) \end{array} \quad (2)$$

$$E(t_r|F_r=h) = E(t_h|F_h=0,R_h=j,F_r=d_r+3) \quad (3)$$

$$E(t_r|F_r=d_r+1) = E(t_r|F_r=d_r) \quad (4)$$

$$E(t_r|F_r=d_r+3) = 0 \quad (5)$$

$$E(t_r) = \min_{h=max(f,1)\ldots d_r+2} E(t_r|F_r=f) \quad (6)$$

where:

- $E(t_r|F_r=i)$: Expected time to complete the task if the system performs the next attempt, given i preceding failures.
- $P(S|F=i)$: Probability of completing the task, given i preceding failures.
- $E(t|S,F=i)$: Expected value of the distribution formed by all data points in which the task was completed with i preceding failures.
- F: Number of preceding failures.
- h: Number of failures after which control will pass to the operator.
- R_h: Number of previously observed human runs.
- d: Max number of preceding failures for which data exists.
- j: Current number of previously observed human runs.
- f: Current number of preceding failures.

As can be seen from Equation 2, the expected time to complete the task if the autonomous system performs the next attempt is a recursive sum, representing the successive attempts made after a failure to complete the task. Equation 4 permits the autonomous system to make an attempt with one more preceding failure than has been previously observed. As we can see from Equation 6, the final value for $E(t_r)$ is chosen by determining the proper point in the future to hand control to the human ($h >= 1$ because $E(t_r)$ represents the assignment of the next attempt to the autonomous system). Equation 5 prevents infinite mutual recursion, since the human's user model includes passing control to the autonomous system (see Equation 9).

We introduce two new elements in the human's model: the learning curve and the possibility of the autonomous system requesting control. If the operator is inexperienced, it is inaccurate to model her performance as a sample from a static distribution. Rather, she is still learning, and it is more appropriate to model $E(t_h)$ by predicting the next point on the learning curve, rather than simply taking the expected value of a distribution of past data. This learning curve (Equation 7) is a logarithmic curve fitted to the available data. We have conducted a series of experiments, discussed below, to determine a reasonable model of $L(x)$ and how best to use it as a predictor of the human's performance in the next trial. Equation 8 represents the system's belief that the human has failed if they are taking too long to complete a task. This is necessary to detect operator failure, since the human operator rarely, if ever, voluntarily gives up. Additional factors may play a role in $E(t_h)$, such as current workload and fatigue. However, they would likely play a relatively straightforward additive or multiplicative role in Equation 7, and are thus neglected for now.

$$N(t) = L(x_{t+1}|x_{1\ldots t}) \tag{7}$$

$$M(s,i) = P(s|F_h = i) - P(t_h > cN(R_h)|s, F_h = i) \tag{8}$$

$$\begin{aligned} &E(t_h|F_h = i, R_h = j, F_r = k) = \\ &\quad M(S_h, F_h)N(R_h) \\ &\quad + M(\neg S_h, F_h)\,(N(R_h) + E(t_h|F_h = i+1, R_h = j+1, F_r = k)) \\ &\quad + P(t_h > cN(R_h))E(t_r|F_r = k) \end{aligned} \tag{9}$$

where:

$N(t)$: Predicted time to complete the task based on a learning curve L fitted to all prior observations.

$L(x_{t+1}|x_{1\ldots t})$: The value of a fitted learning curve for trial $t+1$, given t prior observations.

$M(s,i)$: The probability of $s \in \{S, \neg S\}$ given i preceding failures, less the probability that the autonomous system will request control, given s and i.

c: A constant which determines the time when the system believes the human has failed and the time when it will request control.

$P(t_h > cN(R_h) \; |s, F_h = i)$: The probability that the human will take more than c times the expected time to complete the task.

$E\left(t_h| \begin{array}{r} F_h = i, \\ R_h = j, \\ F_r = k \end{array}\right)$: Expected time to complete the task if the human performs the i'th attempt, with j historical attempts, and k preceding failures by the autonomous system.

The Implemented Model. When implementing any method, tradeoffs must be made between the fidelity of the model and the difficulty involved in estimating the variables involved. For our initial implementation, we set $P(fcat_h) = P(fcat_r) = 0$, $price(h) = price(r_a)$, and $price(r_t) = 0$, collapsing Equation 1 to a straightforward comparison of $E(t_r)$ and $E(t_h)$. The model of the autonomous system was implemented as described in Equations 2 - 6. However, we chose to simplify the human model by disregarding all factors affecting the prediction of the human's performance except previous observations and an estimate of the human's learning curve. We also set $c = \infty$ to prevent the system from requesting control. Since no subject ever failed to complete the task,

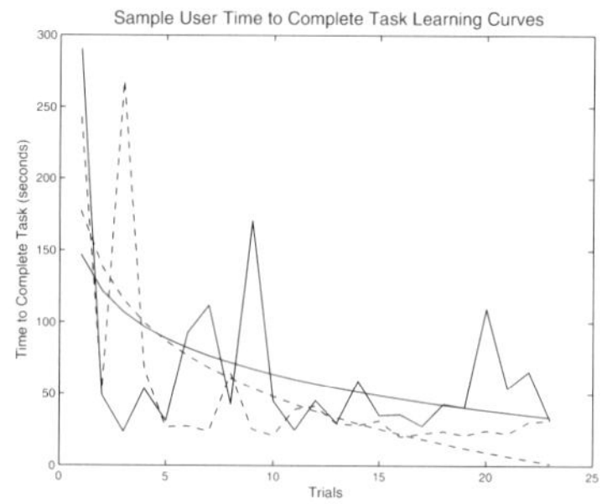

Figure 2. The raw data and fitted logarithmic learning curves for two sample subjects - Subject A's raw data and fitted curve are plotted as solid lines, while Subject B's are dashed lines.

$P(S_h|F_h=0)=1.0)$, resulting in Equation 9 collapsing to $E(t_h)=L(x_{t+1}|x_{1...t})$, or simply the prediction of her time to complete the task based on her learning curve and prior experience. The resulting simplified calculation directly compares the system's and human's estimated time to complete a task in order to assign responsibility.

4.3 Results

In order to build our initial model of the human learning curve $L(x)$ within our domain, we conducted a series of experiments with eight subjects to assess and model their learning curves for direct control of the mobile manipulator's arm. The goal was to develop an understanding of the repeatability a human's time to complete a task; how many trials it would take to eliminate learning effects; and whether a single learning function could be accurately parameterized on a per-user basis, to allow the system to attempt to predict the performance of a user who had not been fully trained.

In order to focus purely on the skill of controlling the arm and minimize confounding variables, the task consisted of docking one end of a beam to a node while directly observing the workspace and controlling the arm via a SpaceMouse (a six degree of freedom input device). Data from a representative two of our eight subjects can be found in Figure 2 — this data contains roughly an average amount of noise. As can be seen, the raw data is quite noisy, with large deviations between successive runs. However, it does consistently trend downwards, and while examining all eight data sets, we discovered that a logarithmic learning curve of the form $L(x)=a*ln(x)+b$, with the parameters a and b fitted to each user's data, yielded a more predictive fit than linear, exponential, or quadratic models. On average, 10 trials worth of data were necessary for the parameters to settle to their final values, but the logarithmic model proved some predictive worth as early as trial 3. Most subjects' performance had plateaued by trial 14.

Taking this into account, we have extended our sliding autonomy system to include the simplified user model described in Section 4.2 for making principled decisions about changes to the team's autonomy levels. The model tracks each operator (and the autonomous system) for every task in the scenario. Given the instability of the initial few trials for most subjects, the model merely predicts the average of the past observed trials for $E(t_h)$ until three trials worth of data are accumulated. Between trials three and fourteen, a logarithmic curve is fit to the data and is used to predict $E(t_h)$ on the next instance of the task. After trial fourteen, the average of the past three trials (discarding outliers) is used, since most subjects' performance plateaus by this point, with the occasional outlier. This allows the autonomous system to make appropriate use of even an inexperienced operator.

5. Future Work

A variety of opportunities to expand upon this work exist. Our simplified model needs to be verified in our assembly system and potentially refined to provide satisfactory predictions. The model could also be extended to extrapolate performance on unobserved tasks from performance on different, but related, tasks. Knowledge of upcoming tasks could also be incorporated into the model, allowing the system to make locally inefficient decisions in order to train a novice user to provide future benefits. Similarly, if the human fatigues over the course of a scenario, the system could avoid asking for help when the human only provides marginal benefit, in order to keep her rested for tasks where she is orders of magnitude better.

6. Conclusion

We have formulated a method for making principled decisions about when to involve a remote human operator in a multi-agent assembly task. We conducted initial user experiments, determining that a parameterized logarithmic function provides an adequate fit to users' observed learning curves. Such a function, tuned to each user as data is observed, provides a usable predictive model of their future performance. Combined with our predictive model of autonomous system performance, this simplified model has been implemented within our sliding autonomy system, allowing the system to make principled decisions about when to request assistance from the operator.

Acknowledgments

The authors would like to thank the many individuals who contributed to the DIRA and Trestle projects over the years: Rob Ambrose, David Apfelbaum,

Jon Brookshire, Rob Burridge, Brad Hamner, Dave Hershberger, Myung Hwangbo, Simon Mehalek, Metrica/TRACLabs, Josue Ramos, Trey Smith, Pete Staritz, and Paul Tompkins.

References

Brookshire, J., Singh, S., and Simmons, R. (2004). Preliminary results in sliding autonomy for assembly by coordinated teams. In *Proceedings of International Conference on Intelligent Robots and Systems (IROS) 2004*.

Fleming, M. and Cohen, R. (2001). A user modeling approach to determining system initiative in mixed-initiative ai systems. In *Proceedings of User Modeling (UM) 2001*.

Fong, T., Thorpe, C., and Baur, C. (2003). Robot, asker of questions. *Robotics and Autonomous systems*, 42.

Goldberg, D., Cicirello, V., Dias, M., Simmons, R., Smith, S., and Stentz, A. (2003). Market-based multi-robot planning in a distributed layered architecture. In *Multi-Robot Systems: From Swarms to Intelligent Automata: Proceedings from the 2003 International Workshop on Multi-Robot Systems*, volume 2, pages 27–38. Kluwer Academic Publishers.

Kortenkamp, D., Burridge, R., Bonasso, P., Schrenkenghoist, D., and Hudson, M. (1999). An intelligent osftware architecture for semi-autonomous robot control. In *Autonomy Control Software Workshop, Autonomous Agents 99*.

McGeoch, G. O. and Irion, A. L. (1952). *The Psychology of Human Learning*. New York: Longmans.

Scerri, P. and Pynadath, D. (2002). Towards adjustable autonomy for the real world.

Sellner, B., Simmons, R., and Singh, S. (2005). Syndicate: A decentralized, layered architecture for tightly coordinating heterogeneous teams. In *Submitted to the International Conference on Robotics and Automation (ICRA-05)*.

Simmons, R., Smith, T., Dias, M. B., Goldberg, D., Hershberger, D., Stentz, A., and Zlot, R. (2002). A layered architecture for coordination of mobile robots. In Schultz, A. and Parker, L., editors, *Multi-Robot Systems: From Swarms to Intelligent Automata*. Kluwer.

VII

APPLICATIONS

MULTI-ROBOT CHEMICAL PLUME TRACING

Diana Spears, Dimitri Zarzhitsky
Computer Science Department, University of Wyoming, Laramie WY 82071
dspears,dimzar@cs.uwyo.edu

David Thayer
Department of Physics and Astronomy, University of Wyoming, Laramie WY 82071
drthayer@uwyo.edu

Abstract This paper presents a novel and effective algorithm for the chemical plume tracing (CPT) task, i.e., tracing a toxic chemical gas to its source emitter. This algorithm is based on a firm, theoretical foundation of fluid flow physics. It assumes a team of plume-tracing robots that act as a mobile, distributed, adaptive sensing grid. In addition to presenting the foundation and the algorithm, an empirical comparison is provided between our algorithm and the two most popular alternatives, on a suite of simulated plume configurations of practical interest.

Keywords: Multi-robot, plume tracing, artificial physics

1. Introduction

The objective of this research is the development of an effective, efficient, and robust distributed search-and-identify algorithm for a team of robots that must locate an emitter that is releasing a toxic chemical gas. The basis for this algorithm is a physics-based framework for distributed multi-agent control ((Spears and Gordon, 1999, Spears et al., 2004a)). This framework, called *physicomimetics* or *artificial physics (AP)*, assumes several to hundreds of simple, inexpensive mobile robotic agents with limited processing power and a small set of on-board sensors. Using AP, the robots will configure into geometric lattice formations that are preserved as the robots navigate around obstacles toward a source location ((Spears et al., 2004b)).

In this paper, we present a novel algorithm for chemical plume tracing (CPT) that is built upon the AP framework. The CPT task consists of finding the chemical, tracking the chemical to its source emitter and, finally, identifying the emitter. Our CPT algorithm, called *fluxotaxis*, combines the strengths of the two most popular chemical plume tracing techniques in use today. Further-

L.E. Parker et al. (eds.),
Multi-Robot Systems. From Swarms to Intelligent Automata. Volume III, 211–221.

more, it is founded upon theoretical principles of fluid dynamics. Our algorithm assumes an AP-maintained lattice that acts as a distributed computational fluid dynamics grid for calculating derivatives of *flow-field variables*, such as fluid velocity and chemical concentration. After presenting the algorithm and its theoretical foundation, the paper presents the comparative performance of our algorithm against the most popular alternatives, on a suite of simulated chemical plumes.

2. Motivation

The authors' goal is to design control algorithms that scale well to a large number of robots, ranging perhaps from ten agents to a thousand and beyond. In order to achieve this goal, two things are necessary: a formal theory upon which the algorithm is based, and a suitable task that can be used to test the algorithm. The task of chemical plume tracing has posed problems for a number of years in a variety of manufacturing and military applications. In light of the current national concern with security and the possibility of a chemical terrorist attack, several private and government agencies, including DHS and DARPA, have expressed interest in updating current techniques used to track hazardous plumes, and improving the search strategies used to locate the toxin emitter (e.g., (Board on Atmospheric Sciences and Climate, 2003, Hsu, 2003, Cordesman, 2001, Caldwell et al., 1997)).

Because CPT involves tracing fluid flow, physics is the natural choice for the theoretical foundation of our CPT approach. In particular, the well-studied field of fluid dynamics is suitable for the development and validation of our algorithms.

Another advantage of using a physics-based foundation is that computational fluid mechanics requires computational meshes for sampling and processing of flow-field variable values. The lattice arrangements that emerge naturally from the physicomimetics framework can be used as computational meshes, capable of performing complex computations in real time, with the added benefit of resilience to failure, and ability to adjust when the environment characteristics change. Furthermore, effective robot configurations for CPT serendipitously coincide with efficient configurations for AP. For instance, the construction of hexagonal formations requires the least amount of communication and sensor information within the AP control framework ((Spears et al., 2004a)); at the same time, a hexagonal lattice was shown ((Carlson et al., 2003)) to have superior boundary characteristics for solving a class of fluid mechanics problems relevant to the CPT task.

3. Related Work

The best understood and most widely applied CPT approach is that of *chemotaxis*, which consists of following a local gradient of the chemical concentration, ρ, within a plume ((Sandini et al., 1993, Special Issue, 2002)). While chemotaxis is very simple to perform, it frequently leads to locations of high concentration in the plume that are not the source, such as a corner of a room.

To overcome this problem, another common approach, called *anemotaxis*, has been developed. Once the chemical has been detected, an anemotaxis-driven agent measures the direction of the fluid's velocity, $\vec{V}$, and navigates "upstream" within the plume ((Hayes et al., 2001, Special Issue, 2002)). Although this can be a very effective strategy for some problems, its limitation is that it can lead to a wind source that is not the chemical emitter.

The advantage of our novel approach is that it is built upon these two most popular predecessors, and it outperforms them on a practical suite of plumes (see below). In the next section, the theoretical foundation of our fluxotaxis algorithm is presented.

4. Fluid Physics

Our approach makes use of the methods and concepts developed in the context of computational fluid dynamics, so a brief review of the relevant material will be useful. Flow of fluids is governed by three fundamental laws: the conservation of mass, conservation of momentum (Newton's Second Law), and the conservation of energy ((Anderson, 1995)). Collectively, these equations are known as the *Governing Equations*. These equations come in several forms, but we will focus on a form that is based on the time analysis of a differential volume spatially fixed in the flow field ((Anderson, 1995)). For instance, the simplest equation, the conservation of mass, is:

$$-\frac{\partial \rho}{\partial t} = \nabla \cdot (\rho \vec{V}) \tag{1}$$

Here, ρ denotes the mass density of the chemical, $\vec{V}$ is the fluid's velocity (ρ and $\vec{V}$ are flow-field variables), and t denotes time. For any real flow of practical interest, an analytical solution of the Governing Equations is impossible to obtain, due to the inherent non-linearity of the fluid dynamic systems. Thus, one computational approach replaces the continuous partial derivatives with the corresponding discretized finite-difference approximations, and computes the unknown flow-field variables using a computational grid which spans the region of interest. Our algorithm takes advantage of the lattice formations formed by our robotic agents to simulate the computational grid, thereby allowing the agents to perform a sophisticated (but computationally efficient) analysis of the flow and make navigational decisions based on this analysis.

Computational efficiency of the fluxotaxis algorithm derives from local sensing and communication. Robots determine the range and bearing to their immediate, detected neighbors in order to calculate virtual forces that hold the robots together in a lattice formation.[1] The robots also share sensed chemical density and wind velocity values with these neighbors in order to decide the next direction to move for plume tracing. These flow-field variable values are mapped (using standard coordinate transformations) to the robots' own local coordinate axes. Based on these values, each robot independently decides the best direction to move to trace the plume. This direction is translated into a virtual force. Finally, each robot takes the resultant vector of the plume-following force and the lattice-preserving force – this resultant vector determines the robot's next move. A balance of forces results in smooth movement of the lattice as a whole toward the plume source, with avoidance of obstacles along the way using virtual repulsion by the obstacles.

5. The Fluxotaxis Algorithm

The product $\rho\vec{V}$ is called the *mass flux* ((Anderson, 1995)), and represents the time rate of change of mass flow per unit area; dimensional analysis shows that $\rho\vec{V}$ is simply mass/(area · time). This flux formally combines the ρ focus of chemotaxis with the $\vec{V}$ focus of anemotaxis, and it is the basis of our *fluxo*taxis algorithm.

For the CPT task, our simulations mimic real-world laboratory robot conditions as faithfully as possible. The simulated robot grid consists of six robots in a hexagon formation with one additional robot in the middle. [2] Each robot in the grid serves as both a sensor and a decision unit. Complex actions are modeled as taking a realistically long time to complete, while the flow continues to evolve. Robots' maximum speed corresponds to hardware constraints. Note that at any given time, the hexagonal robot lattice will have a specific radius, which depends on the inter-robot distances. Initial and maximum lattice radii are given realistic values, based on collision avoidance and communication range constraints.

Fluxotaxis addresses all three subtasks of the CPT task: finding the chemical, tracing the plume and, finally, identifying the emitter. We describe the emitter identification subtask first, because it follows directly from the theory just presented.

5.1 Emitter Identification Subtask

The RHS of (1) represents the divergence of mass flux within the differential volume. Divergence is a convenient way to quantify the change of a vector field in space. Although our approach is applicable to 3D geometries, for greater

Algorithm: `anemotaxis`

```
while TRUE do
  ensure lattice radius and location are within limits;
  interrupt current lattice action if action time limit expired;
  if lattice is within plume and sensors detect V flow direction
    then execute move_upstream()
    else execute cast()
  end if
end while
```

Strategy: `move_upstream`

```
  average the direction of fluid flow V across the lattice;
  move opposite the flow V direction for 1 step at maximum speed
```

Figure 1. The anemotaxis algorithm.

simplicity, we express the mass flux divergence in 2D Cartesian coordinates as

$$\nabla\cdot(\rho\vec{V}) = u\frac{\partial\rho}{\partial x} + \rho\frac{\partial u}{\partial x} + v\frac{\partial\rho}{\partial y} + \rho\frac{\partial v}{\partial y} \tag{2}$$

where $\vec{V} = u\hat{i} + v\hat{j}$ and $\hat{i}$ and $\hat{j}$ are unit vectors in the x and y coordinate directions, respectively. (Recall that each robot has its own local coordinate system.) If at some spatial point location P, $\nabla\cdot(\rho\vec{V}) > 0$, then it is said that point P is a source of $\rho\vec{V}$, while $\nabla\cdot(\rho\vec{V}) < 0$ indicates a sink of $\rho\vec{V}$. Recall that the mass flux, $\rho\vec{V}$, represents the time rate of change of mass flow per unit area. The role of mass flux in the CPT task can be better understood with the aid of the Divergence Theorem from vector calculus ((Hughes-Hallett et al., 1998)):

$$\int_W \nabla\cdot(\rho\vec{V})\mathrm{d}W = \oint_S(\rho\vec{V})\cdot\mathrm{d}S \tag{3}$$

This equation, where W is the control volume and S is the bounding surface of the volume, allows us to formally define the intuitive notion that a control volume containing a source will have a positive mass flux divergence, while a control volume containing a sink will have a negative mass flux divergence. This result serves as our basic criterion for theoretically identifying a chemical emitter, which is a source. In particular, (3) shows that if the robots encircle a suspected emitter, and the total mass flux exiting the circle of robots consistently exceeds some small, empirically-determined threshold, then the robots are known to surround a true chemical emitter. To the best of our knowledge, previous criteria for emitter identification are purely heuristic, e.g., ((Special Issue, 2002)). Ours is the first with a firm theoretical basis.

5.2 Finding the Chemical

Prior to tracing a chemical to the source emitter, the chemical plume must first be located. The most common method for doing this is called *casting*

Algorithm: `chemotaxis`

while TRUE **do**
ensure lattice radius and location are within limits;
interrupt current lattice action if action time limit expired;
if lattice is within plume
then execute `move_to_max_density()`
else execute `cast()`
end if
end while

Strategy: `move_to_max_density`
take the sensor reading of ρ across the lattice;
move to the location of the maximum ρ

Figure 2. The chemotaxis algorithm.

and typically consists of a zigzag or spiraling motion to increase exploration of the region ((Hayes et al., 2001)). In this research, we extend the traditional casting approach to improve its effectiveness. We are able to do this because unlike prior approaches we have a lattice of several or more, rather than just one or two, robots. In addition to translational motion, the improved casting technique also includes lattice expansions and contractions (implemented using AP forces with local information). For the comparisons presented below, all three algorithms (chemotaxis, anemotaxis, and fluxotaxis) use a lattice with this improved method of casting.

5.3 Tracing the Plume

The algorithm implementations for anemotaxis and chemotaxis are direct from the literature ((Special Issue, 2002)) other than the casting modification described above. Figures 1–3 show the three CPT algorithms, which are composed of low-level strategies.

Low-level functions ensure that the lattice never expands beyond the maximum allowed radius and environmental boundaries, and that the agents never travel faster than the maximum speed. Also of importance is the fact that anemotaxis and chemotaxis algorithms lack stopping criteria ((Special Issue, 2002)). Therefore, for the sake of fair comparisons, we decided not to have fluxotaxis stop either, even when its emitter identification procedure succeeds.

6. The Plume Simulator

The majority of our research so far has been in simulation. In addition to designing and writing our own chemical plume simulations, we have obtained simulations and plume data from other sources. The experimental results reported in this paper use the most practical and well-developed of all the simulators that we have worked with, developed by Farrell et al. (2002) at the University of California ((Special Issue, 2002)).[3] Farrell's simulator is

```
Algorithm: fluxotaxis
    while emitter identification test fails do
        ensure lattice radius and location are within limits;
        interrupt current lattice action if action time limit expired;
        if lattice is within plume
          then
            if more than 50% of total ρ is sensed by the center agent
              then contract the lattice to minimal radius
              else execute chem_region()
            end if
        else execute cast()
        end if
    end while

Strategy: chem_region
        sense total lattice ρ over 3 different lattice radii;
        compute ρ centroid C_i where i ∈ RADIUS_{inner,middle,outer}
        if ρ increases with each radial increase
          then move to the centroid of the centroids C_i
        else if outermost ρ is greater than innermost
          then move to the location of the C_outer centroid
        else if ρ decreases with each increasing radius
          then execute flux_ring()
        else execute cast()
        end if

Strategy: flux_ring
        compute the maximum incoming flux, ρV⃗, for 3 different lattice radii;
        if maximum influx exceeds a flux threshold
          then move to the location of the maximum incoming flux, ρV⃗
        else
            compute the maximum outgoing flux, ρV⃗;
            if maximum outflux exceeds flux threshold
              then move to the location of the maximum outgoing flux
              else execute cast()
```

Figure 3. The fluxotaxis algorithm.

especially well-suited to our needs because it is: computationally efficient, realistic (i.e., its transient and statistical behavior are very close to measured actual plumes), and it is multi-scale – including both molecular diffusion of the chemical and advective transportation by wind movement. Rather than a continuous, time-averaged model, Farrell's simulator models the plume as a filament-based emission of chemical "puffs."

7. Experiments

7.1 Design

In this section, we compare chemotaxis, anemotaxis and fluxotaxis on a suite of simulated, realistic plume scenarios. Thirty five different scenarios/ plumes were chosen, each containing a dynamic plume evolving over a 100 sq. ft. region, without obstacles. The plumes in our suite fall into the categories

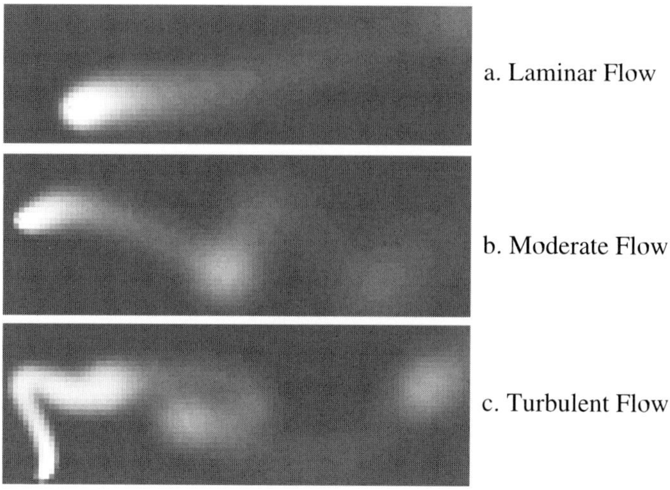

Figure 4. The three general types of flow and plume configurations simulated for the CPT experiments.

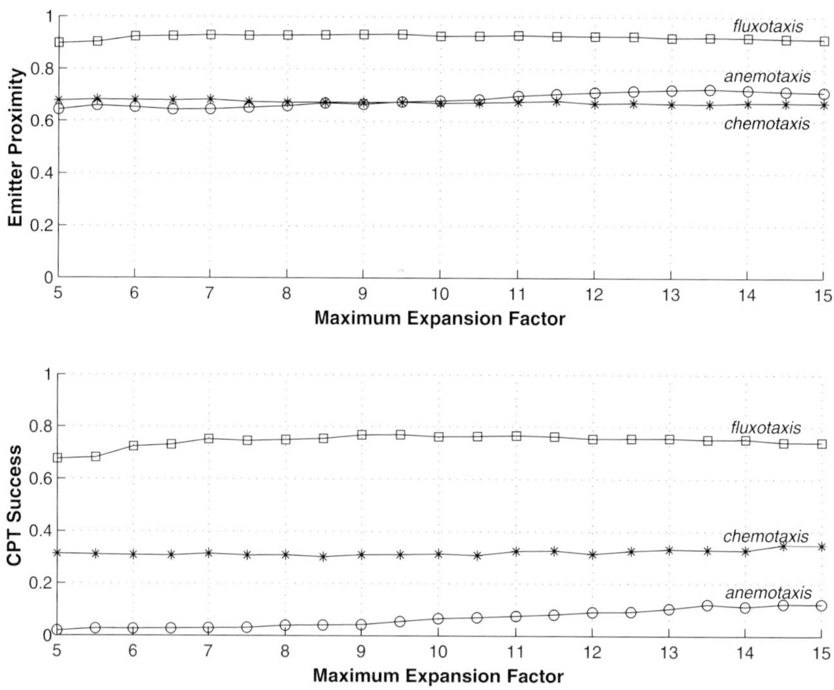

Figure 5. Performance of the three CPT algorithms with respect to varying expansion factors over 35 plumes with 200 random starting locations.

Table 1. CPT algorithm performance averaged over 35 plumes and all parameter variations, showing mean ± standard deviation.

Algorithm	*Proximity*	*Success*
Anemotaxis	0.6843 ± 0.0291	0.0667 ± 0.0362
Chemotaxis	0.6745 ± 0.0049	0.3184 ± 0.0132
Fluxotaxis	0.9235 ± 0.0089	0.7460 ± 0.0250

laminar, moderately turbulent (transitional), and turbulent (see Figure 4). An agent is assumed to have a dimension of one sq. ft. and a speed of six inches per second. The lattice movement as a whole is determined using way points spaced three feet apart. We varied the maximum lattice expansion factor, because it appears to have the largest impact on performance of any of the system parameters. For each plume and expansion factor parameter setting, there were 200 independent simulation runs, each with a different, randomly-chosen starting location of the lattice. All algorithms were compared over identical runs (e.g., identical starting locations and all other parameters the same). Each run terminated after 5000 steps ($\sim$ 80 minutes real time). The time limit was determined empirically as being sufficient time to find the plume with casting and then get to the emitter. Note that the plume evolution and lattice movement are concurrent, as in a real plume, and the lattice is driven by the resultant vector of virtual formation and plume-tracing forces.

Because neither anemotaxis nor chemotaxis techniques specify a control scheme for the lattice radius, but fluxotaxis does, in order to keep the evaluation as fair as possible, anemotaxis and chemotaxis driven lattices were allowed to expand or contract their radii at random. The initial size of the lattice radius was fixed at 1.5 feet. The maximum expansion factor was set at 15, resulting in a maximum lattice diameter of 45 feet.

The evaluation metric consisted of two components: *proximity* of the center agent to the true location of the chemical emitter, and a Boolean measure of emitter containment (by the lattice) at termination, which is called a *success*. Success for chemotaxis and anemotaxis were determined by a global observer algorithm, used only for performance evaluation. Note that the second metric is influenced by the maximum radius; a larger radius implies a higher likelihood of success. Results presented next show the averages over all lattices and parameter settings, with 200 independent runs per lattice and setting.

7.2 Results

The graphs of the experimental results are shown in Figure 5. These graphs show average performance over all plumes with respect to the maximum expansion factor. Note that in the graph of proximity, with a higher maximum expansion factor, anemotaxis outperforms chemotaxis. This is due to an oscillation in anemotaxis at large lattice radii when the lattice moves upwind to get closer to the emitter, overshoots it and loses the chemical, switches to casting, and then back to upwind-following behavior. In the graph plotting success rate, all algorithms improve when the radius is larger because of the increased likelihood of surrounding the emitter in this case. Finally, statistics averaged over all the measures and runs are given in Table 1. A Wilcoxon rank sum test was performed, which showed a statistically significant advantage of fluxotaxis over both chemotaxis and anemotaxis, on both performance metrics, at a $p < 0.05$ level of significance.

7.3 Conclusions

From Figure 5 and Table 1 we can see that the relative advantage of fluxotaxis over the alternative algorithms is clear, both in terms of proximity *and* success rate. In conclusion, for our chosen performance metrics, fluxotaxis substantially outperforms the leading alternative CPT algorithms on our suite of simulated plumes. Its success is due to the fact that it is a synergistic, theoretically-founded combination of the two alternatives. Note that our performance metric does not account for the time it takes to get to the emitter although, if it did, fluxotaxis would be the winner in this respect also, based on all of our experiments so far (not reported here). Future experiments will focus on determining how the algorithms scale when obstacles are introduced into the environment.

8. Summary and Future Work

Further algorithm development in simulation will include online learning of thresholds, maintaining a history of observations, and modeling of sensor characteristics (e.g., size, number, noise). An important near-term focus will be on porting the simulation to actual robots in a laboratory plume emission setting. The University of Wyoming Distributed Robotics Laboratory has a team of several small robots that have successfully demonstrated robot self-organization into lattices, obstacle avoidance, and goal seeking in formation ((Spears et al., 2004b)). The next step is to integrate chemical sensors with the robot processors and test the fluxotaxis algorithm with emissions of volatile organic compound gases. For the long-term goal of this project, we plan to transition to outdoor robots, including heterogeneous teams of ground-based and micro-air vehicle platforms.

Acknowledgments

The authors would like to thank Bill Spears for numerous useful comments and suggestions.

Notes

1. The forces are based on a Lennard-Jones potential.

2. More robots would translate to better performance, but we model our laboratory, which has seven robots.

3. We actually use a re-implementation of the original simulator from Farrell. Our re-implementation adds efficiency and generality to the original code.

References

Anderson, J. D. (1995). *Computational Fluid Dynamics*. McGraw-Hill, Inc.

Board on Atmospheric Sciences and Climate (2003). *Tracking and Predicting the Atmospheric Dispersion of Hazardous Material Releases: Implications for Homeland Security*. National Academies Press.

Caldwell, S. L., D'Agostino, D. M., McGeary, R. A., Purdy, H. L., Schwartz, M. J., Weeter, G. K., and Wyrsch, R. J. (1997). Combating terrorism: Federal agencies' efforts to implement national policy and strategy. Congressional report GAO/NSIAD-97-254, produced by the United States General Accounting Office.

Carlson, E. S., Sun, H., Smith, D. H., and Zhang, J. (2003). *Third Order Accuracy of the 4-Point Hexagonal Net Grid. Finite Difference Scheme for Solving the 2D Helmholtz Equation.* Technical Report No. 379-03, Department of Computer Science, University of Kentucky.

Cordesman, A. H. (2001). Defending America: Asymmetric and terrorist attacks with chemical weapons. Report produced by the Center for Strategic and International Studies (CSIS).

Hayes, A., Martinoli, A., and Goodman, R. (2001). Swarm robotic odor localization. In *Proceedings of the IEEE/RSJ International Conference on Intelligent Robots and Systems (IROS'01)*.

Hsu, S. S. (2003). Sensors may track terror's fallout. *Washington Post*, page A01.

Hughes-Hallett, D., Gleason, A. M., and et al., W. M. (1998). *Calculus: Single and Multivariable*. John Wiley and Sons.

Sandini, G., Lucarini, G., and Varoli, M. (1993). Gradient driven self-organizing systems. In *Proceedings of the IEEE/RSJ International Conference on Intelligent Robots and Systems (IROS'93)*.

Spears, W. and Gordon, D. (1999). Using artificial physics to control agents. In *Proceedings of the IEEE Conference on Information, Intelligence, and Systems (ICIIS'99)*.

Spears, W., Heil, R., Spears, D., and Zarzhitsky, D. (2004a). Physicomimetics for mobile robot formations. In *Proceedings of the Third International Joint Conference on Autonomous Agents and Multi Agent Systems (AAMAS-04)*, volume 3, pages 1528–1529.

Spears, W., Spears, D., Hamann, J., and Heil, R. (2004b). Distributed, physics-based control of swarms of vehicles. In *Autonomous Robots*, volume 17(2-3).

Special Issue (2002). Chemical plume tracing. In Cowen, E., editor, *Environmental Fluid Mechanics*, volume 2. Kluwer.

DEPLOYING AIR-GROUND MULTI-ROBOT TEAMS IN URBAN ENVIRONMENTS

L. Chaimowicz, A. Cowley, D. Gomez-Ibanez, B. Grocholsky, M. A. Hsieh, H. Hsu, J. F. Keller, V. Kumar, R. Swaminathan, C. J. Taylor
GRASP Laboratory – University of Pennsylvania
Philadelphia – PA – USA

Abstract We present some of the work performed in the GRASP Laboratory with the objective of deploying multi-robot teams in urban environments. Specifically, we focus on three important issues in this type of mission: the development of tools for providing situational awareness, the use of air and ground vehicles for cooperative sensing and the construction of radio maps to keep team connectivity. We describe the main approaches that we have been using for tackling these issues and present some preliminary results from experiments conducted with our team of air and ground vehicles.

Keywords: Air-ground cooperation, situational awareness, radio mapping.

1. Introduction

Urban environments provide unique challenges for the deployment of multi-robot teams. In this type of environment, buildings pose 3-D constraints on visibility, communication network performance is difficult to predict and GPS measurements can be unreliable or even unavailable. We believe that a network of aerial and ground vehicles working in cooperation can have a better performance in these type of environments. By constructing a three-dimensional sensing network, teams of air and ground vehicles can obtain better and more complete information and thus be more robust to the challenges posed by these environments. For this, it is necessary to keep the network tightly integrated at all times since vehicles have to support each other in order to function with synergy. Also, it is important to provide ways for a human operator to command the whole network, and not individual vehicles.

In this paper, we present some of the efforts that have been done by the GRASP Laboratory – University of Pennsylvania for deploying teams of air and ground vehicles in urban environments as part of the MARS2020 project.

L.E. Parker et al. (eds.),
Multi-Robot Systems. From Swarms to Intelligent Automata. Volume III, 223–234.

Sponsored by DARPA, this project is focused on the development of critical technologies required to realize network-centric control of heterogeneous platforms that is strategically responsive, survivable and sustainable for reconnaissance, surveillance or search and rescue type missions. In this endeavor, the University of Pennsylvania is teamed with the Georgia Institute of Technology, the University of Southern California and BBN Technologies.

At the University of Pennsylvania, our MARS2020 research thrust is to establish the overall paradigm, modeling framework and the software architecture to enable a minimum number of human operators to manage a heterogeneous robotic team with varying degrees of autonomy. The central features of our approach are to organize the robotic platforms for network centric autonomous operations; to develop small team communication-sensitive behaviors, which allow robots to perform alongside humans as full team members; to enable the team to learn and adapt to changing terrain conditions that may impact communication network performance and localization information (e.g., GPS, line of sight sensing, etc.); and to develop computationally distributed strategies to provide an unprecedented level of situational awareness. The effort will result in an integrated team of UAVs and UGVs, in which the team and the network will adapt to the needs and commands of a remotely located human operator to provide situational awareness.

This paper is organized as follows: the next section presents our multi-robot team. Section 3 describes some of the basic capabilities of our team in terms of localizing and navigating in urban terrains. Sections 4 to 6 discuss our main research thrusts, namely: situational awareness, air-ground cooperation and cooperative radio mapping. Finally, section 7 brings the conclusion of this paper.

2. Hardware and Software Testbed

Our multi-robot team consists of 5 unmanned ground vehicles (UGVs), 2 fixed wing aircraft and a blimp (Figure 1). The UGVs employ a commercially available, radio controlled scale model truck with suspension and chassis modified for autonomous duty. The chassis is approximately 480 mm long and 350 mm high. Mounted in the center of the chassis is a Pentium III laptop computer. Each UGV contains a specially designed Universal Serial Bus (USB) device which controls drive motors, odometry, steering servos and a camera pan mount with input from the PC. A GPS receiver is mounted on the top of an antenna tower, and an inertial measurement unit (IMU) is mounted between the rear wheels. A forward-looking stereo camera pair is mounted on a pan mount which can pivot 180 degrees to look left and right. A small embedded computer with 802.11 wireless Ethernet handles network connectivity. This junction box (JBox), developed jointly by the Space and Naval Warfare

Systems Center, BBN Technologies, and the GRASP Lab, handles multi-hop routing in an ad-hoc wireless network. The UGV is powered by one or two hot swappable lithium polymer battery packs, each with 50 watt hour capacity.

The fixed wing aircraft are quarter scale Piper Cub model aircraft equipped with the Piccolo autopilot by Cloud Cap Technology. The autopilot provides innerloop attitude and velocity stabilization control allowing research to focus on guidance for mission level tasks. In addition to the sensors within the autopilot, the air vehicles carry a sensor pod containing a high resolution firewire camera, inertial sensors and a 10Hz GPS receiver. A spread-spectrum radio modem is used for communications between air vehicles and the operator base station. Ground based system components communicate through an Ad-Hoc 802.11b network. We also have a medium sized blimp (9 meter length) that has nearly a 3 kg payload for research equipment. It is equipped a GPS, an inertial measurement unit capable of sensing rates and attitudes, a video camera that can be slewed in azimuth and elevation, and the onboard computing and communication hardware to be autonomous but also capable of being dynamically redirected by a remote human operator.

Figure 1. Multi-robot team composed of air and ground vehicles.

We are using ROCI (Remote Object Control Interface) (Chaimowicz et al., 2003, Cowley et al., 2004) for programming and tasking both the ground and air vehicles. ROCI is a high level operating system useful for programming and managing networks of robots and sensors. In ROCI, each robot is considered a node which contains several processing and sensing modules and may export different types of services and data to other nodes. Each node runs a kernel that mediates the interactions of the robots in a team. The kernel is responsible for handling program allocation and injection, managing the network and maintaining an updated database of other nodes in the ROCI network. The control functionality needed by such a kernel is made possible by self-contained, reusable modules. Each module encapsulates a process which acts on data available on its inputs and presents its results on well defined outputs. Thus, complex tasks can be built by connecting inputs and outputs of specific modules. A ROCI task is a way of describing an instance of a collection of ROCI modules to be run on a single node, and how they interact at runtime. It is defined in an XML file which specifies the modules that are needed to achieve the goal, any necessary module-specific parameters, and the connections between these modules. At runtime, these connections are made through a pin architecture that provides a strongly typed, network transparent communication framework.

3. Localization and Navigation

Two of the key requirements for the robots is the ability to localize themselves and navigate in urban environments. A Kalman filter framework is employed to estimate robot localization. Prediction is driven by wheel encoder odometry and inertial measurements from a low cost IMU. Appropriate observation models allow various sources of position information to be incorporated. These include on-board GPS, robot observations from external vision sensors and landmarks observed by the on-board camera.

Our robots navigate based on a list of desired waypoints. Each waypoint is a pair of georeferenced coordinates that specify a destination point for the robot. The waypoints can be specified manually through a user interface or can be input directly to the navigation module by other modules and tasks. Sometimes it may be necessary to automatically generate a list of intermediary waypoints between the robot current position and the desired destination, so that the robots will follow an specific path to their goal. One way of doing this is to create a graph based on a Voronoi Diagram of the environment and use it as a roadmap for planing the intermediary waypoints. The Voronoi Diagram can be generated beforehand, using overhead imagery obtained by the air vehicles. Another possibility is to use "mission scripts", which will be discussed in the next section.

A trajectory controller generates linear and angular velocities for the robot based on its current position and the next desired waypoint. A robot considers a waypoint reached when the distance between them is less than a threshold ε. Local obstacle avoidance is accomplished through the use of the robot's stereo vision system. Images captured simultaneously from the two cameras are used to generate a medium-density depth map through a multi-pass process of confidence adjustment. This depth map is converted to a two-dimensional occupancy grid centered on one of the cameras in the stereo setup. Several trajectory arcs, corresponding to various vehicle movements (e.g. turn left, go straight, or turn right) can be compared against this grid to verify if a collision would result. The use of a finite number of discrete trajectories, rather than a more complete shortest path solver, lets the system run at the rate of the camera with very little processor overhead.

Figure 2 shows the trajectory performed by a robot following a sequence of waypoints in the MOUT (Military Operations on Urban Terrain) site at Fort Benning, the main test site for this project. The MOUT site is a replica of a small city consisting of 17 two and three store buildings, streets and access roads. It is configured with cameras that allow a multiple view tracking of training missions. It also features a small airfield, making it a suitable test ground for air-ground cooperation.

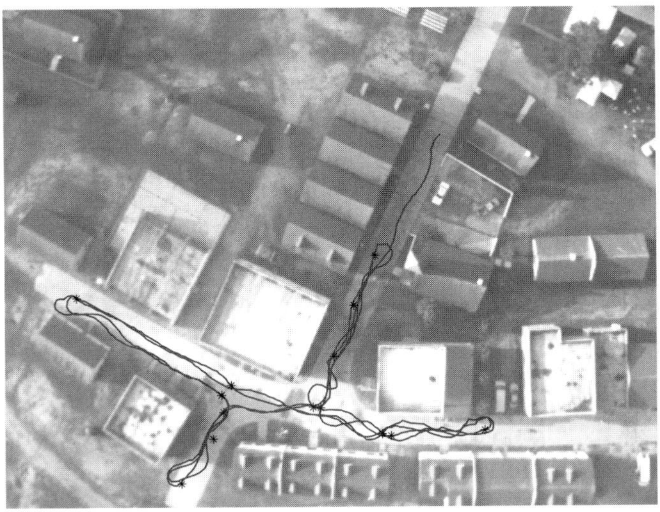

Figure 2. Trajectory of a robot navigating using a sequence of waypoints. The waypoints are depicted as '*' and the robot executes the sequence twice.

4. Situational Awareness

In our framework, the main interface between a human operator and the robot team is the ROCI Browser. The browser displays the multi-robot network hierarchically: the human operator can browse nodes on the network, tasks running on each node, the modules that make up each task, and pins within those modules. The browser's main job is to give a user command and control over the network as well the ability to retrieve and visualize information from any one of the distributed nodes.

Using the browser, the user can start and stop the execution of tasks in the robots remotely, change task parameters or send relevant control information for the robots. Elaborated missions can be constructed using *scripts*. Mission scripts can be generated online or offline, and specify a sequence of actions that should be performed by a team member. For example, capturing panoramic images at different waypoints, or navigating through multiple intermediate waypoints before reaching a target site. A synchronization mechanism allows for coordinated efforts between multiple robots, and a success/failure condition check on the outcome of each action makes limited branching possible. We are currently utilizing these capabilities to support a multi-robot signal strength mapping mission with intelligent recovery behavior if at any point any of the robots lose radio connectivity to the other members. This specific mission will be discussed in Section 6.

The visualization and exploitation of data generated by the multi-robot team is one of the main features of our situational awareness framework. To access and visualize the data, a human operator interacts with the ROCI Browser, part of which is a display generation runtime made up of a collection of plug-in Display Modules that convert incoming pin data to raster images. Data can be retrieved from different nodes equipped with various types of sensors such as GPS receivers, IMU readers, cameras, etc, and can be combined to give the user a rich view of the mission.

Figure 3 shows a snapshot of the browser during the execution of one of our experiments in Ft. Benning. It displays an image of the MOUT site taken previously from one of the UAVs overlayed with different information acquired from the robots. For example, the thick lines represent signal strength of the radio connection between the five nodes in our network. One robot can be seen in the center of the image surrounded by a octagon that indicates the uncertainty of its localization. As mention in Section 3 the robot position is estimated by fusing information from several sources, in this case an on board GPS, odometry, and an external overhead camera. If desired, these individual observations can also be displayed on the browser.

Both users and autonomous agents also have the ability to access sensor data through a distributed database infrastructure. The foundation of this system is made up of instances of a Logger module that is capable of receiving, indexing, and storing any type of pin data. Using these modules, a user can instrument a task with loggers listening to any pin communications. This data is primarily indexed by time (an index that is globally meaningful in the context of distributed sensing), but can also be indexed by more sensor-specific methods (such as position). By using shared indices to join multiple logs in an on-demand fashion, human operators and programmed agents are able to make use of data as it becomes available on the network without relying on any pre-defined schemas. A query architecture is used to interact with this distributed database. Active queries, made up of executable code, are sent to the nodes that are storing the relevant information in order to minimize the amount

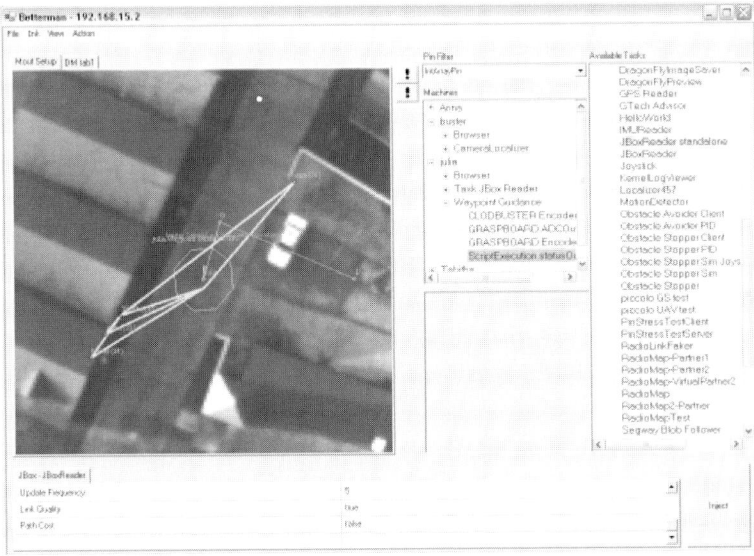

Figure 3. Snapshot of the ROCI Browser during the execution of a mission. Overlayed in the aerial picture, the thick lines represent radio connectivity between nodes and the octagon shows the uncertainty in the localization of one of the robots.

of data that must be transferred over the network. In other words, the query is moved to the data, rather than the other way around. This system helps to eliminate the distinction between low-level sensor data and high-level fused structures by removing the need to hard code every type of useful structure.

5. Air-Ground Cooperation

The use of air and ground robotic vehicles working in cooperation can be very important in tasks involving the reconnaissance and exploration of cluttered urban environments where communication and GPS localization may be unreliable. In this type of mission, groups of unmanned air vehicles (UAVs) could significatively help the ground vehicles (UGVs) by providing localization data and acting as communication relays.

One of our first experiments in air-ground cooperation was to try to localize our ground vehicles using a sequence of images taken from the blimp. Basically, for each image we had to compute the projection matrix M that relates the position of the robot in a global coordinate frame with its pixel coordinates in the image. We compared two complementary approaches for computing M: the first used a set of known landmarks in the image while the second relied on the measurements from the blimp's on board sensors (GPS/IMU) and the intrinsic parameters of the camera. The localization results were compared with measurements from a GPS on board of the ground vehicle.

As discussed in (Chaimowicz et al., 2004), these experiments demonstrated that none of these approaches could be applied alone if we need a localization system that is applicable, reliable, and accurate. In spite of being very precise, the air-ground localization using known landmarks can not always be applied because it requires the identification of a certain number of world locations to register the image, which is impractical in general situations. On the other hand, the approach based on the blimp's on board sensors did not performed well due to the combination of different sensor errors without an adequate fusion methodology. These preliminary results motivated us to pursue more sophisticated methods for performing the cooperative localization. The general idea is to fuse information from different sources in a systematic way in order to have a more reliable and accurate localization system.

Therefore, we developed a related approach in which air and ground vehicles can collaborate to localize ground features. As noted, when detecting ground features from images taken from a blimp or an airplane, their exact location on the ground is always subject to errors in attitude and location estimates. Thus, for robust localization of ground targets, it is imperative to know and reduce the uncertainty in their position.

This approach builds on previous endeavors in decentralized data fusion (DDF) (Manyika and Durrant-Whyte, 1994). DDF provides a decentralized estimation framework equivalent to the linearized Kalman filter. Decentralized active sensor networks (Grocholsky et al., 2003) extend this to include a control layer that refines the quality of the estimates obtained. The established architecture and methodology is used here.

As detailed in (Grocholsky et al., 2004), our methodology combines UAV and UGV ground feature observations, actively deploying the sensor platforms to maximize the quality of the location estimates obtained. The different perspective provided to sensors on-board air and ground vehicles results in significantly different sensing accuracy and coverage qualities. A collaborative feature search and localization example is illustrated in Figure 4. The approach exploits the complementary character of UAV and UGV sensor platforms. The UAV rapidly covers the designated search area, detecting features and providing relatively uncertain location estimates. UGVs deploy to refine the feature location estimates. Localization accuracy beyond that achievable by UAV sensing alone is realized without requiring the UGVs to conduct an extensive area search.

6. Cooperative Radio-Mapping

Communication is essential for coordination in most cooperative control and sensing paradigms. Successful deployment of multi-robot mapping and exploration, surveillance, and search and rescue relies in large part on a reliable

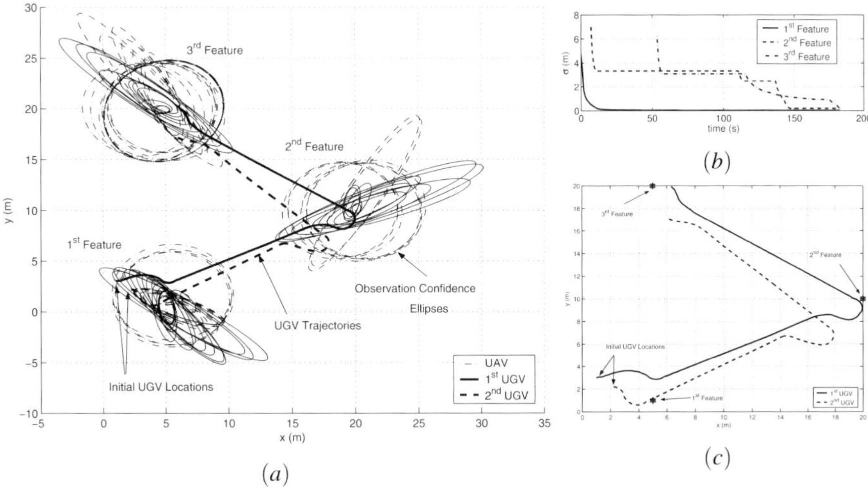

Figure 4. Details of the ground feature localization process. (*a*) Confidence ellipses associated with UAV and UGV feature observations. Indicating the significant uncertainty reduction with UGV sensing distance. (*b*) The estimate standard deviation over time. Constant values indicate the time between UAV spotting and UGV refinement. (*c*) UGV trajectories and true feature locations.

communication network. Often times, these tasks are executed in environments that are adverse to wireless communications. Radio propagation characteristics are difficult to a predict a priori since they depend upon a variety of factors such as transmission power, terrain characteristics, 3-D geometry of the environment, and interference from other sources (Neskovic et al., 2000). The difficulty in accurately modeling radio propagation in urban environments makes it difficult to design multi-agent systems such that the individual agents operate within reliable communication range at all times. In this section, we consider the problem of acquiring information for radio connectivity maps in urban terrains that can be used to plan multi-robot tasks and also serve as useful perceptual information.

A radio connectivity map is a function that returns the signal strength between any two positions in the environment. The radio connectivity construction can be formulated as a graph exploration problem for small teams of robots. An overhead surveillance picture is used to automatically generate roadmaps for motion planning and determine a plan to obtain the desired signal strength measurements (Hsieh et al., 2004). The plan consists of a list of waypoints for each robot such that radio signal strength measurements for the connectivity map is obtained as team members simultaneously traverse through each of their respective waypoints.

Radio connectivity maps can be therefore used to plan multi-robot tasks to increase the probability of a reliable communication network during the execution phase. This however will not guarantee that radio signal strength measurements obtained during the exploration and execution phases will not differ due to the sensitivity of radio propagation to environmental factors. Therefore, to ensure reliable communication during task execution, additional recovery behaviors such as returning to the last position the robot was able to successfully communicate with other team members should be included. Ideally, the measurements obtained during the exploration phase can be used to construct a limited model for radio propagation in the given environment such that when coupled with additional reactive behaviors, a reliable communication network can be maintained during deployment.

Preliminary experiments were performed using our ground vehicles to test the radio connectivity at the Ft. Benning MOUT site. In these experiments, each robot is individually tasked with the corresponding list of waypoints determined by the algorithm described in (Hsieh et al., 2004). Team members navigate to their designated waypoints and broadcasts an "arrival" message. Once the robots have completed the radio signal strength measurements, they broadcast a "ready to go" to notify each other to move on to their next targeted location. This is repeated until every member has traversed through all the waypoints on their list. The waypoints are selected to minimize the probability of losing connectivity under line-of-sight conditions in the planning phase to ensure the success of the synchronization based on line-of-sight propagation characteristics that can be determined a priori. Figure 6 shows some measurements of the radio signal strength between pairs of positions in the environment. An edge between two pairs of positions shows that the signal strength between the two locations is above the desired threshold. The weights on the edges (barely visible) denote the signal strength that was measured between the two locations. In these experiments, the signal strength was measured using the JBox, described in Section 2.

In the future, we would like to incorporate more reactive behaviors so as to be able to do some on-line mapping instead of collecting data for a set of completely predetermined locations. Furthermore, it is often the case that the exploration of the radio map of the scene is being carried out concurrently with other activities such as environmental monitoring or situational awareness. Thus, another area which we plan to address is pursuing the radio mapping with other objectives and which must be effectively balanced against the other mission goals.

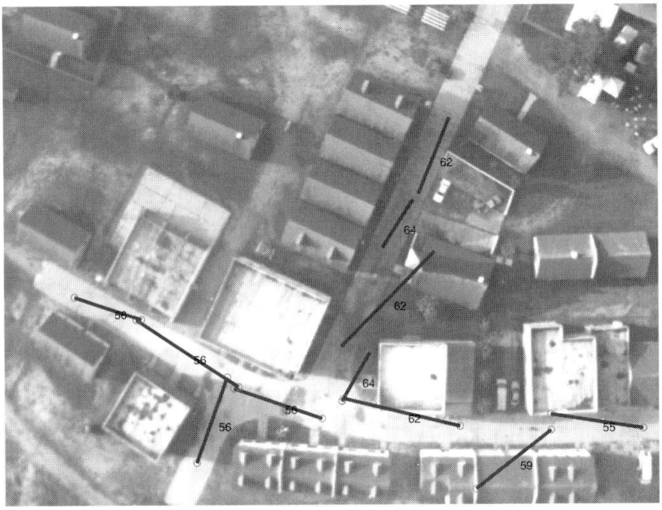

Figure 5. Preliminary experimental radio connectivity map for the MOUT site obtained using our multi-robot testbed.

7. Conclusions

This paper presented some of the work that has been done in the GRASP Lab. as part of the DARPA-MARS2020 project for deploying teams of robots in urban environments. We introduced our hardware and software framework, discussed some important issues related to situational awareness, air-ground cooperation and cooperative radio mapping, and presented some preliminary results obtained during field tests. This project is scheduled to culminate with a demonstration of the performance of the integrated team of UAVs, UGVs, and a human leader in a reconnaissance type mission at the Fort Benning McKenna Range MOUT site in December 2004.

Acknowledgments

This work was in part supported by The Defense Advanced Research Projects Agency (DARPA MARS NBCH1020012).

References

Chaimowicz, L., Cowley, A., Sabella, V., and Taylor, C. J. (2003). ROCI: A distributed framework for multi-robot perception and control. In *Proceedings of the 2003 IEEE/RJS International Conference on Intelligent Robots and Systems*, pages 266–271.

Chaimowicz, L., Grocholsky, B., Keller, J. F., Kumar, V., and Taylor, C. J. (2004). Experiments in multirobot air-ground coordination. In *Proceedings of the 2004 IEEE International Conference on Robotics and Automation*, pages 4053–4058.

Cowley, A., Hsu, H., and Taylor, C. (2004). Distributed sensor databases for multi-robot teams. In *Proceedings of the 2004 IEEE International Conference on Robotics and Automation*.

Grocholsky, B., Bayraktar, S., Kumar, V., and Pappas, G. (2004). Uav and ugv collaboration for active ground feature search and localization. In *Proc. of the AIAA 3rd "Unmanned Unlimited" Technical Conference*.

Grocholsky, B., Makarenko, A., Kaupp, T., and Durrant-Whyte, H. (2003). Scalable control of decentralised sensor platforms. In *Information Processing in Sensor Networks: 2nd Int Workshop, IPSN03*, pages 96–112.

Hsieh, M. A., Kumar, V., and Taylor, C. J. (2004). Constructing radio signal strength maps with multiple robots. In *Proceedings of the 2004 IEEE International Conference on Robotics and Automation*, pages 4184–4189.

Manyika, J. and Durrant-Whyte, H. (1994). *Data Fusion and Sensor Management: An Information-Theoretic Approach*. Prentice Hall.

Neskovic, A., Neskovic, N., and Paunovic, G. (2000). Modern approaches in modeling of mobile radio systems propagation environment. *IEEE Communications Surveys*.

PRECISION MANIPULATION WITH COOPERATIVE ROBOTS

Ashley Stroupe, Terry Huntsberger, Avi Okon, Hrand Aghazarian

Jet Propulsion Laboratory, California Institute of Technology
4800 Oak Grove Drive, Pasadena, CA, 91109
{Ashley.Stroupe, Terry.Huntsberger, Avi.Okon, Hrand.Aghazarian}@jpl.nasa.gov

Abstract: We present a cooperative approach to robotic precision manipulation tasks in the context of autonomous robotic construction. Precision manipulation requires a firm grasp, which constraints the team to rigidly maintain formation during transport and manipulation. A leader/follower approach with force sensing to provide relative formation information and vision to provide team position relative to construction components is applied. Our approach demonstrates successful, reliable performance of a construction task requiring cooperative transport and placement of structure components. Qualitative and quantitative performance results are provided.

Keywords: Cooperative Robots, Cooperative Transport, Robot Construction

1. Introduction

The 2004 NASA vision for space exploration calls for a sustainable presence in space, beginning with a human return to the Moon in 2020 (NASA, 2004). A sustainable robotic or human presence requires deploying and maintaining infrastructure to provide power, living quarters, and resource acquisition and utilization; deployment must be autonomous for safety and reliability. Space operation places constraints on rover power, computing, communication, and mass. JPL's Planetary Robotics Lab (PRL) is developing autonomous technologies to perform construction related tasks under these constraints. One focus area is cooperative transport and precision manipulation of large rigid components over natural terrain using

L.E. Parker et al. (eds.),
Multi-Robot Systems. From Swarms to Intelligent Automata. Volume III, 235–248.

fused sensor information from both robots. We present details and quantitative results of our approach, an extension of previous work (Stroupe, et al., 2004) and done under the CAMPOUT architecture (Huntsberger, et al., 2003).

This work addresses several challenges of cooperative transport and precision manipulation. Precision manipulation requires a rigid grasp, which places a hard constraint on the relative rover formation that must be accommodated, even though the rovers cannot directly observe their relative poses. Additionally, rovers must jointly select appropriate actions based on all available sensor information. Lastly, rovers cannot act on independent sensor information, but must fuse information to move jointly; the methods for fusing information must be determined.

2. Background and Related Work

Precision manipulation of large components requires rovers to have a rigid grasp and therefore to remain in a fixed relative formation, even if rovers cannot directly obtain each other's relative position. Cooperative transport has primarily focused on pushing on flat floors, typically relying on direct observation or communication of relative position (Parker 1994, Qingguo and Payandeh, 2003; Rus, et al., 1995). In some cases, robots use forces imparted by the object to communicate, but robots adjust position to obtain desired forces (Brown and Jennings, 1995). These approaches are not applicable to precision manipulation in three-dimensions, rigid formations, or operation in natural terrain. Most precision cooperative manipulation using force feedback for implicit communication is for fixed-base manipulators (Mukaiyama, et al., 1996), and is not applicable for construction. Formation following typically uses potential fields, which require observation or communication of relative pose and do not accommodate hard constraints (Balch and Arkin, 1998; Carpin and Parker, 2002; Desay, et al., 1999). Most similar to our task is Omnimate (Borenstein, 2000), which uses a compliant linkage between two mobile platforms; Omnimate compensates for uneven floors and moderate wheel-slippage by controlling wheel velocity based on observed angular difference between the expected and observed lines of contact between the platforms. To date, cooperative transport in rigid formation is limited to JPL's Robot Work Crew (Huntsberger, et al. 2004, Trebi-Ollenu, et al., 2002), and mobile precision manipulation has not yet been demonstrated.

3. Cooperative Transport / Manipulation

3.1 Task Domain and Description

Two rovers, heterogeneous in size and arm configuration, are equipped with a forward-facing stereo camera pair and a three-axis force-torque sensor on the arm's wrist. The team can cooperatively carry long beams that are stacked and interlocked if positioned accurately. Each beam has a grasping point at either end; the gripper lower finger passes through the grasp point and the upper fingers close on the top of the beam. Rovers obtain beam location from the stereo vision of three fiducials on each end. The test environment is a PRL sand pit that simulates natural terrain.

Figure 1. Left: RCC Rovers SRR (left) and SRR2K (right) are heterogeneous; note SRR2K's gripper is not in line the rover body center. Right: Structure composed of stacked beams; note each beam is alternately offset laterally from the one it rests upon. The inset shows the grasping point, the fiducials and the interlocking cones on one beam.

Tasks are in the context of building a four-sided structure by stacking beams. The team must cooperatively transport beams from storage to the structure and then align at the structure while remaining in formation to prevent dangerous forces on rover arms. Once aligned, the team must precisely place the beam in the structure while keeping arms at appropriate relative positions. Robots return to storage and repeat the process to build the structure. Current work demonstrates end-to-end acquisition and placement of a single beam; results here focus on the cooperative aspects.

3.2 Cooperative Behaviors

3.2.1 Aligning with the Structure

Rovers position themselves relative to the structure based on stereo vision of the fiducials on the beams already in the structure. To place a beam into the structure, the rover arms must maintain separation; the rigid grasp cannot accommodate arbitrary offsets in rover position. Thus, team alignment with the structure must be precise (within 1 cm). Slippage, errors in drive kinematics, and errors in estimated structure location lead to alignment errors. Vision errors are minimized by validating observations against the model of fiducials on the beam and repeating observations as necessary. Additionally, the vision is calibrated relative to the arm to offset any kinematics errors in the arm control. The team reports any unrecoverable errors to allow corrective intervention.

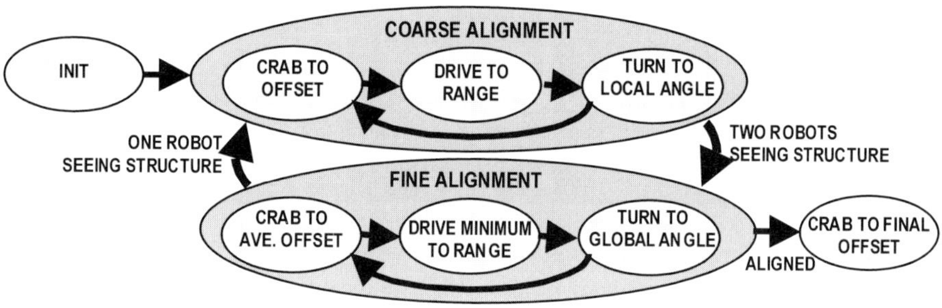

Figure 2. Simplified finite state machine representing the alignment process, with different modes for action selection depending on the amount of data available.

To reduce the effects of errors, the algorithm uses an iterative process of adjusting range, lateral offset, and heading, as illustrated in the state machine of *Figure 2* and the diagrams of *Figure 3*. To ensure rovers are executing parallel motions, they share data and the leader selects an action and sends to the follower; roles can be dynamically changed. As position is refined, visual information becomes more accurate and errors from slip in previous motions are corrected. If an iteration completes without corrections, the team is properly aligned. To maximize visual accuracy, the team iteratively aligns directly in front of structure fiducials first, and then iteratively aligns the lateral offset appropriate for placing the beam. The leader selects an action (movement magnitude and direction) using fused information from both robots, if both see the structure, or the individual information from either robot, if only one robot sees the structure.

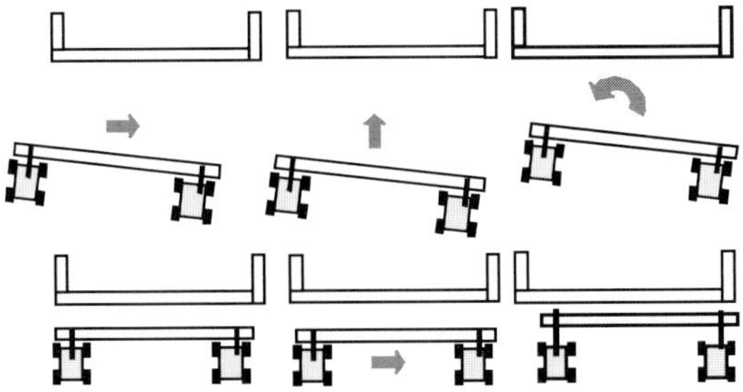

Figure 3. Top: Iterative alignment process of crab (left), drive (center), and Ackermann turn (right). *Bottom*: Aligned team (left) crabs to offset beam for placement (center) and extend arms to place beam (right).

When only one robot sees the structure, the team must move the team so as to allow the other rover to also see the structure. This involves attempting to position the other rover at an appropriate range and lateral position. Due to the linear formation constraint imposed by the beam, if one rover is approximately aligned, the other rover should be near alignment and able to see the structure. Thus, the observing rover approximately aligns itself to bring the other rover into position. Driving (forward or backward) and crabbing (lateral driving) will bring the observing rover close to desired range (X_D) and zero arm lateral offset by moving the team those distances; this is shown in (1), where (X, Y) is the observed structure position and M_Y is the manipulator arm lateral offset. To bring the team parallel, the rovers execute an Ackermann based on the observing rover's estimate of its heading relative to the structure; this is determined using the difference in range to the fiducials on the left and the fiducials on the right (ΔX_T) and the known lateral distance between fiducials (ΔY_T). The rover turns to counter this angle as in (2).

$$D = \begin{matrix} X - X_D \\ Y - M_Y \end{matrix} \quad (1)$$

$$A = -\tan^{-1}(\Delta X_T, \Delta Y_T) \quad (2)$$

$$A = \tan^{-1}(\Delta X, \Delta Y_G) \quad (3)$$

When both robots see the structure (during fine alignment and during the final lateral offset), information is fused differently for different conditions and different motions. Driving brings one member of the team to the correct range. The team selects to execute the shortest drive (X) computed by (1) to prevent overshooting desired range. Crabbing aligns the rovers laterally with the structure. The team executes a crab in magnitude to satisfy both rovers: the average offset. If the rovers are in a linear formation and aligned with the structure, the average Y offset from (1) will bring both rovers into position. To bring the rovers to the same range (aligning the team and beam parallel to the structure), an Ackermann turn is done. The turn is computed as in (3), using the difference in range (ΔX) for the two robots and the distance between beam grasping points (ΔY_G); the further robot drives forward while the closer robot drives backward, equalizing the distance. There is one exception: if in the previous step only one rover saw the structure and the team anticipated turning based on local rather than global information, the follower's range (required for determining global angle) is not available, but instead the follower's local angle was shared. In this case, the turn is the average of both local angles from (2) to approximately satisfy both. Rather than attempting to robustly identify this situation on both rovers and resend the proper data, this approach saves additional communication and ensures the rovers stay in parallel states. Turns too small to execute (less than one degree) are approximated by a drive, with each rover driving independently to the correct range from (1); these small offsets can be tolerated in forward driving.

If both robots lose sight of the structure, they move forward in increments of 5 cm until at least one regains observation or until a timeout is reached.

3.2.2 Driving in Rigid Formation

As described in 3.2.1, the team performs three motions in formation: drive, crab, and Ackermann turn. To keep the beam straight and prevent large stresses on the arms, the team must remain at the desired lateral separation and in-line with each other. Keeping formation allows the team to achieve the correct relative position to place the component in the structure. The arms can only accommodate small lateral offsets during component placement without inducing large forces, thus accurate rover positioning is essential. By implementing motions determined by the leader (rather than in a distributed manner), cooperative moves are always identical. Synchronizing cooperative moves reduces (but does not eliminate) errors due to time offsets. Leader-controlled motion and synchronization do not mitigate initial formation errors.

Empirical data has indicated correlations between torque about the vertical axis and rover alignment and between force along the horizontal axis and rover separation; the team detects errors in relative formation using forces and torques imparted through the beam. Once a formation error is detected it can be corrected as shown in *Figure 4*. By allowing the follower to adjust its velocity, it can eliminate formation errors. For driving, the follower speeds up if torque indicates it is behind and slows down if it is ahead, *Figure 4* illustrates the case where the follower is on the right. The relationship between torque and force and velocity correction is dependent on direction of motion. Currently, only velocity control is used to adjust formation; steering adjustments to account for off-axis formation errors is in development. To ensure forces and torques do not build up before velocity control can compensate, the follower starts first (the leader delays 2 seconds after synchronization). To prevent overreaction to sensor noise, the sensor is sampled at 20 times the action selection rate and averaged.

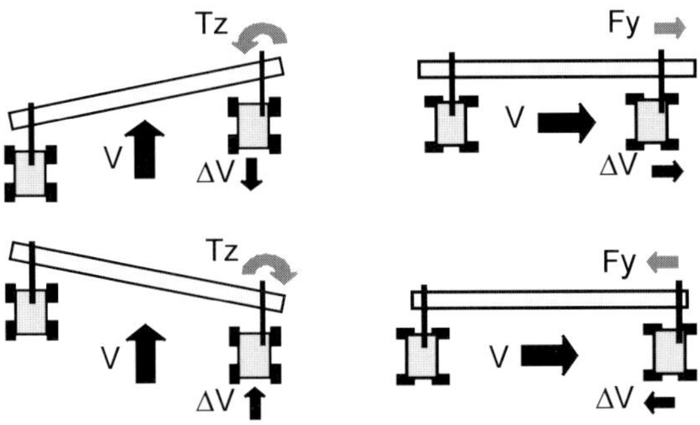

Figure 4. Left: Follower adjusts velocity to eliminate torque. The follower slows down to reduce counterclockwise torque and recover alignment (top) or speeds up to reduce clockwise torque and recover linear alignment (bottom). Right: Follower adjusts crabbing velocity to eliminate force. Follower speeds up to increase spacing and reduce positive force (top) or slows to decrease spacing and reduce negative force (bottom).

The velocity controllers are PI controllers designed to maximize response time, minimize oscillation (particularly at the end of moves), and stop both rovers upon failure. Parameters for control were empirically determined based on observed performance. (4) shows the general controller, where V_B is the base velocity, E is the error term, ΣE is total accumulated error, and ΣE_{MAX} is the magnitude limit on the accumulated error. (5) shows drive velocity error, where T_Z is torque about the vertical axis T_{Z0} is the reference torque about the vertical axis, T_{FY} is torque about the vertical axis due to lateral forces rather than angular misalignment of the beam. Drive control parameters are $K_P = 0.4$, $K_I = 0.0$, and $\Sigma E_{MAX} = 0.0$. (6) shows the error for crabbing velocity, where F_Y is lateral force and F_{Y0} is reference lateral force. Control parameters for crabbing are $K_P = 0.6$, $K_I = 0.05$, and $\Sigma E_{MAX} = 6.0$. The single-step change in velocity is limited by ΔV_{MAX} as in (7).

$$V_{NEW} = V_B + K_P E + K_I \Sigma E \tag{4}$$

$$E_D = T_Z - T_{Z0} - T_{FY} \tag{5}$$

$$E_C = F_Y - F_{Y0} \tag{6}$$

$$V = \begin{cases} V_{LAST} + \Delta V_{MAX} & \text{if } V_{NEW} - V_{LAST} > \Delta V_{MAX} \\ V_{LAST} - \Delta V_{MAX} & \text{if } V_{NEW} - V_{LAST} < \Delta V_{MAX} \\ V_{NEW} & \text{otherwise} \end{cases} \tag{7}$$

While only the follower uses force-torque feedback to control velocity, both rovers monitor forces and torques. If either rover detects a sustained force or torque larger above a threshold, failure is detected and the rover stops driving. This quickly increases force and torque on the other (still moving) robot past threshold, so that it also detects a failure and stops. These thresholds are set based on empirical data. The maximum allowed torque magnitude about the vertical axis is 4.0 N-m and the maximum allowed lateral force magnitude is 44.6 N.

4. Experimental Results

4.1 Cooperative Transport with Adaptive Velocity

To demonstrate the team's ability to remain in formation and keep arm forces and torques within bounds, a series of tests were performed. For these experiments, the rovers start in formation with the follower on the right. Forces and torques minimized and the reference torques and forces are set to the initial values. A drive or crab command is sent and behavior is observed. Between repetitions of an experiment, the rovers are repositioned to reduce forces and torques, but the reference force/torque is not reset; this investigates ability to accommodate small initial offsets in formation. Force-torque profiles with velocity control and fixed velocity are compared.

Table 1 and *Table 2* summarize performance. The resulting mean and standard deviation of the force or torque during nominal driving is reported (from after initial reaction time to initiation of deceleration).

Table 1. Forward Drive Torques (N-m)

Distance	Velocity Control	Fixed Velocity, No Delay	Fixed Velocity, Delay
300 cm	-0.04 ± 0.16	-0.43 ± 0.25	-2.09 ± 1.02
	-0.05 ± 0.22	-0.26 ± 0.43	-2.22 ±0.80
	-0.04 ± 0.21	-0.95 ± 0.43	Failure
	0.02 ± 0.20	-0.06 ± 0.69	-2.03 ± 0.79
	-0.03 ± 0.20	-1.76 ± 0.79	-2.30 ± 0.81

Table 2. Crab Drive Forces (N)

Distance	Velocity Control	Fixed Velocity, No Delay	Fixed Velocity, Delay
80 cm	-3.76 ± 7.16	3.91 ± 8.67	Failed
	0.05 ± 5.38	-1.48 ± 8.67	-14.04 ± 7.59
	-0.44 ± 4.59	-5.54 ± 6.66	-7.30 ± 9.42
-80 cm	1.61 ± 6.15	-6.73 ± 12.38	5.27 ± 9.90
	-0.02 ± 3.93	6.98 ± 11.06	Failure
	-0.78 ± 4.45	15.56 ± 9.86	-3.56 ±9.34

Results show velocity control keeps forces and torques low (within bounds) and with low variance despite initial offsets; fixed velocity shows higher torques (approximately 19 times higher mean) with a high bias and high variance. Fixed velocity may fail, particularly with start offsets, and does fail or nears failure in nearly half of the crabbing experiments. Torque profiles for drives with and without velocity control are compared in *Figure 5*. Torque remains near zero and has little variance while under adaptive velocity control despite initial errors, while torque has large bias and large variance without velocity control. By moving first, the follower can immediately respond to any change in forces and torques.

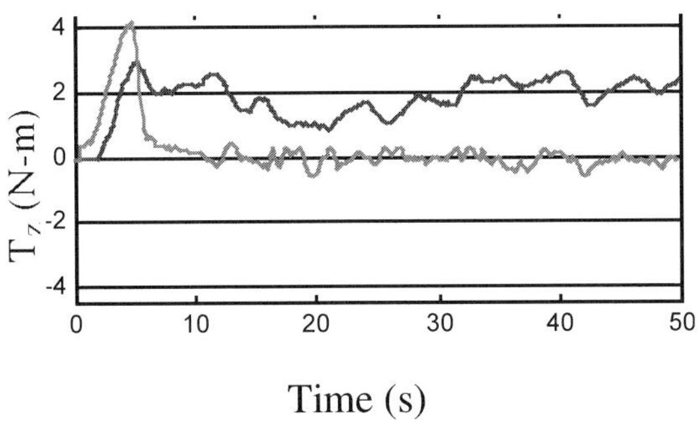

Figure 5. Torque profiles of a 300 cm drive with velocity control and a delay (solid) and without velocity control and no delay (dotted). Note high torques (near failure), large variance, and large bias without control.

Crabbing demonstrates similar results (*Figure 6*); velocity control keeps forces near zero with low variance but fixed velocity nears failure and has high variance.

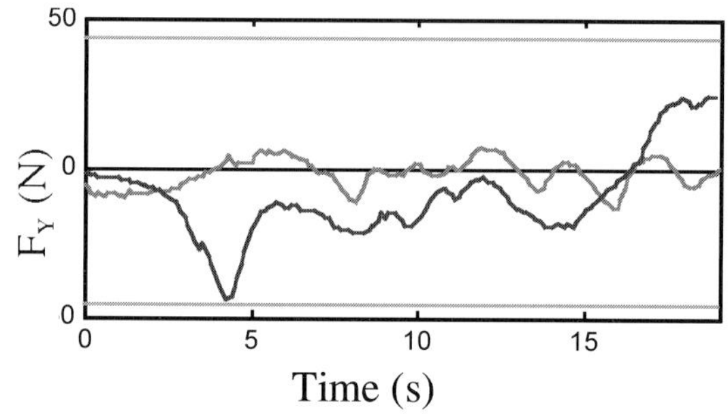

Figure 6. Force profiles of an -80 cm crab with the follower on the right. Note high forces and large variance without control (dotted) compared to velocity control (solid). Force bounds shown dashed.

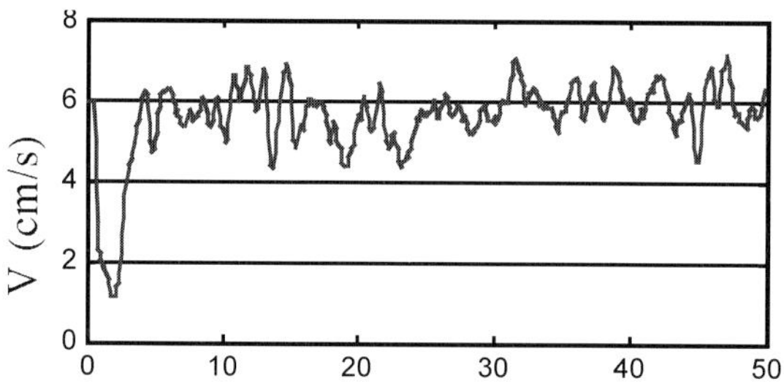

Figure 7. Velocity profile of a 300 cm drive with velocity control. Note that the velocity slows to zero while the follower waits for the leader to catch up, then increases to nominal and remains near nominal for the remainder.

Figure 7 provides a velocity profile for a drive. The follower starts to drive before the leader, leading to a torque. This drops the velocity near zero until the leader begins to catch up. Then, the follower gradually increases speed until the torque is minimized and velocity reaches nominal. Once at nominal velocity, the velocity oscillates slightly compensating for small changes in force due to slip and variation in leader velocity. A crab velocity profile, with similar results, is shown in *Figure 8*.

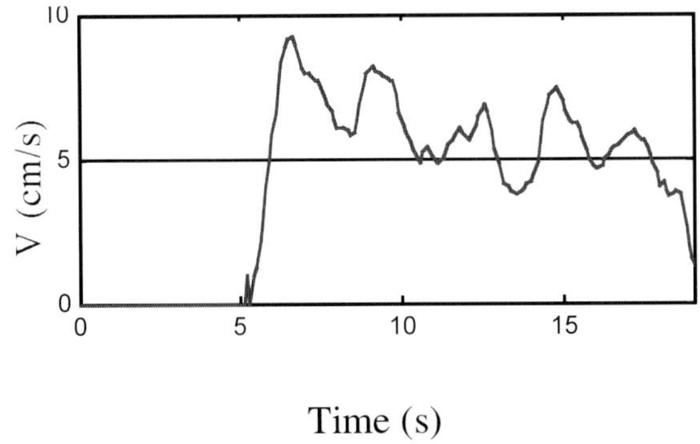

Figure 8. Velocity profile of a -80 cm crab with velocity control. As for drive, the velocity slows to zero while the follower waits for the leader to start moving away, then increases to and remains near nominal until stopping.

No failures of formation occurred during cooperative transport experiments. By using velocity control and ensuring the control begins before forces and torques build up due to partner motion, the rigid formation following is very robust and maintains safe forces on the manipulator arms.

4.2 Cooperative Alignment and Beam Placement

The goal of maintaining formation is to allow the team to successfully and reliably perform the desired task: placement of the beam component into the structure. To demonstrate the reliability of our approach, the team repeated alignment and beam placement multiple times. Any failure was recorded, including failures due to high arm forces or torques (formation failure), failure to align accurately enough for proper beam placement (alignment failure), and failure to place a beam properly due to an unrecognized poor alignment by rover or arm (placement failure). The team is positioned such that at least one rover can see the structure and is placed in formation with minimal forces and torques at a variable relative angle to the structure. Reference forces and torques are set to the initial condition.

Table 3 shows number of alignment attempts listed and number of failures by type. The first series of experiments were run using a previous alignment procedure that did not autonomously correct in the case of both robots simultaneously losing sight of the structure, and coarse alignment (only one robot seeing the structure) used a fixed angle rather than local angle. In the second series, the described approach is used. A series of photographs illustrating an alignment and placement is shown in *Figure 9*. In this example, only SRR can initially see the structure, and the team proceeds through coarse alignment crab, drive, and turn. After one iteration, both rovers can see the structure and the team transitions to fine alignment.

Table 3. Team Alignment and Beam Placement Results

Number of Alignments	Formation Failures	Alignment Failures	Placement Failures
12	0	1	0
5	0	0	0

In the first set there was one failure due to both robots simultaneously losing sight of the structure. No autonomous correction was implemented and the operator corrected the error. There were no failures during the runs of the improved alignment procedure, including a case in which both robots lost sight of the structure and compensated. The iterative process allows correction of errors and multiple checks for validating alignment, making the cooperative transport and manipulation very robust.

Figure 9. Alignment example. Top left: Initially only SRR sees the structure. Top center: After initial crab and drive based on SRR's observations. Top right: Fine alignment in progress after Ackermann turn based on both robots' observations.. Bottom left: Final crab to align for placement. Bottom center: Aligned for placement. Bottom right: Placing beam.

5. Conclusions and Future Work

The Robotic Construction Crew performs cooperative transport and precision manipulation of long rigid beams in the context of a construction task. To place these beams precisely into a structure, the team must align accurately with the structure while remaining in formation. This process iteratively aligns range, offset, and heading. To maintain formation during transport, force-torque feedback is used to adjust velocity. RCC is robust to variable initial conditions, changes in the amount and quality of information available, synchronization errors and temporary communication loss, motion errors due to slippage, and minor driving or arm kinematics errors.

In current and future work, we will refine velocity control for formation following and investigate continual steering adjustments to compensate for formation errors not in the direction of motion. Additionally, force-torque feedback to maintain formations during turns will be included.

Acknowledgments

This work was carried out at the Jet Propulsion Laboratory, California Institute of Technology, under a contract with the National Aeronautics and Space Administration. We thank Paul Schenker, Neville Marzwell and Eric Baumgartner for supporting the RCC effort, and Matthew Robinson for contributing to RCC development. We thank Brett Kennedy, Tony Ganino, Mike Garrett, and Lee Magnone for platform support.

References

Balch, T., and Arkin, R. C. (1998). Behavior-based formation control for multirobot teams. *IEEE Transactions on Robotics and Automation*, 14(6):926-939.

Borenstein, J. (2000). The Omnimate: A guidewire- and beacon-free AGV for highly reconfigurable applications. *International Journal of Production Research*, 38(9):1993-2010.

Brown, R. G. and Jennings, J. S. (1995) A pusher/steerer model for strongly cooperative mobile robot manipulation. *Proceedings of IEEE/RSJ International Conference on Intelligent Robots and Systems*, 3:562-568.

Carpin, S. and Parker, L. E. (2002). Cooperative leader following in a distributed multi-robot system. *Proceedings of IEEE Interernational Conference on Robotics and Automation*, 3:2994-3001.

Desay, J. P., Kumar, V., and Ostrowski, P. Control of Change in Formation for a Team of Mobile Robots. (1999). *Proceedings of IEEE International Conference on Robotics and Automation*, 2:1556-1561.

Huntsberger, T. L., Trebi-Ollennu, A., Aghazarian, H., Schenker, P., Pirjanian, P. and Nayar, H. D. (2004). Distributed Control of Multi-Robot Systems Engaged in Tightly Coupled Tasks. *Autonomous Robots*, 17:79-92.

Huntsberger, T., Pirjanian, P., Trebi-Ollennu, A., Nayar, H. D., Aghazarian, H., Ganino, A., Garrett, M., Joshi, S. S., and Schenker, P.S. (2003). CAMPOUT: A Control Architecture for Tightly Coupled Coordination of Multi-Robot Systems for Planetary Surface Exploration. *IEEE Transactions Systems, Man & Cybernetics, Part A*, 33(5): 550-559.

Mukaiyama, T., Kyunghwan, K., and Hori, Y. (1996). Implementation of cooperative manipulation using decentralized robust position/force control. *Proceedings of the 4th International Workshop on Advanced Motion Control*, 2:529-534.

NASA. (2004). Vision for Space Exploration.

Parker, L. E. (1994). ALLIANCE: an architecture for fault tolerant, cooperative control of heterogeneous mobile robots. *Proceedings of IEEE/RSJ International Conference on Intelligent Robots and Systems*, 2:776-683.

Qingguo, L. and Payandeh, S. (2003). Multi-agent cooperative manipulation with uncertainty: a neural net-based game theoretic approach. *Proceedings of IEEE International Conference on Robotics and Automation*, 3:3607-3612.

Rus, D., Donald, B., and Jennings, J. (1995). Moving furniture with teams of autonomous robots. *Proceedings of IEEE/RSJ Int. Conf. on Intelligent Robots and Systems*, 1:235-242.

Stroupe, A., Huntsberger, T., Okon, A., Aghazarian, H., and Robinson, M. (2005). Behavior-Based Multi-Robot Collaboration for Autonomous Construction Tasks. Submited to *IEEE International Conference on Robotics and Automation.*

Trebi-Ollennu, A., Das, H., Aghazarian, H., Ganino, A., Pirjanian, P., Huntsberger, T., and Schenker, P. (2002). Mars Rover Pair Cooperatively Transporting a Long Payload. *IEEE International Conference on Robotics and Automation.*

VIII

POSTER SHORT PAPERS

A ROBUST MONTE-CARLO ALGORITHM FOR MULTI-ROBOT LOCALIZATION

Vazha Amiranashvili
Computer Science V, RWTH Aachen
Ahornstr. 55, D-52056, Germany
vazha@i5.informatik.rwth-aachen.de

and Gerhard Lakemeyer
Computer Science V, RWTH Aachen
Ahornstr. 55, D-52056, Germany
gerhard@cs.rwth-aachen.de

Abstract This paper presents a new algorithm for the problem of multi-robot localization in a known environment. The approach is based on the mutual refinement by robots of their beliefs about the global poses, whenever they detect each other's paths. The robots try to detect the paths of other robots by comparing their own perception sensor information to that "seen" by other robots. This way the relative poses of the robots can be determined, which in turn enables the exchange of the beliefs about the global poses.

Keywords: Multiple robots, Monte Carlo localization

1. Introduction

Over the last decade a number of very successful probabilistic algorithms has been developed for the solution of the self-localization problem in the single robot case (Fox et al., 1999b, Fox et al., 1999a, Nourbaksh et al., 1995, Simmons and Koenig, 1997, Kaelbling et al., 1996, Burgard et al., 2000, Thrun et al., 2001). All these algorithms estimate posterior distributions over the robot's poses given the sensor data acquired so far by the robot and certain independence assumptions.

It is natural to try to extend the mentioned approaches to the multi-robot case. Indeed, there is a number of works dealing with this problem (Howard et al., 2002, Kurazume and Hirose, 2000, Roumeliotis and Bekey, 2000, Fox et al., 2000). For example in (Roumeliotis and Bekey, 2000) the multi-robot localization is solved by introducing one central Kalman filter for all robots.

L.E. Parker et al. (eds.),
Multi-Robot Systems. From Swarms to Intelligent Automata. Volume III, 251–256.

The method of (Fox et al., 2000) uses the information about the relative poses of the robots to mutually refine the beliefs of the robots about the global poses.

In the approach introduced in this paper we try to solve the s.c. revisiting problem (Fox et al., 2003) in order to determine the relative poses of the robots. However in contrast to that work we deal with a global environment (map) known to all robots.

2. Monte Carlo Localization

Monte Carlo localization (MCL) is a special case of probabilistic state estimation, applied to robotic localization. We assume that robots have odometry and perception sensors (wheel encoders and laser range finders in our case). Our goal is to estimate the conditional probability distribution over all possible poses of the robot given all the sensor data that the robot gathers up to the time of the estimation. Let $l = (x, y, \theta)$ denote a robot location, where x, y and θ are the Cartesian coordinates in the global coordinate system and the orientation (with respect to the x-axis of this system) of a robot respectively. We seek to estimate $bel(l_{0:t}) \triangleq p(l_{0:t}|d_{0:t}, m)$ the belief that at time t the robot has made the path $l_{0:t}$ and its current position is l_t. Here $d_{0:t} = \{a_{0:t-1}, s_{1:t}\}$ is the set of all odometry measurements $a_{0:t-1}$ and perceptions of the environment $s_{1:t}$, while for $i < j$ $x_{i:j} \triangleq \{x_i, x_{i+1}, ..., x_j\}$, m is the map of the environment.

The MCL is an efficient way to compute $bel(l_{0:t})$. It is based on the SIR (sampling/importance resampling) algorithm (Rubin, 1998), which is a version of "particle filters" (Doucet, 1998). It presents the belief function $bel(l_{0:t})$ by a set of weighted samples, or particles. Each particle is a tuple of the form $\langle w, l_{0:t} \rangle$, where $w \geqslant 0$ and if there are N particles in a set, $\sum_{n=1}^{N} w_n = 1$. The particles are then iteratively recomputed till they represent a sufficiently "narrow" distribution of robot poses (Fox et al., 1999a, Fox et al., 1999b, Thrun et al., 2001). Here we assume that the map m is known. On the other hand, in the SLAM problem the map is unknown and should itself be determined. Recently, a Monte Carlo technique called Rao-Blackwellised particle filtering has been applied for the solution of the SLAM problem (Doucet et al., 2000, Montemerlo and Thrun, 2003). Here the map represents a function $\widehat{m}(l_{0:t-1}, s_{1:t-1})$ and in our case $\widehat{m}$ places the laser range scans s_i at the locations l_i, $1 \leqslant i \leqslant t-1$. Later e make use of a simplified version of the mentioned SLAM algorithm to solve the problem of tracking a robot's local pose when the initial global pose of the robot is unknown.

3. Multi-Robot Localization

Our approach to the multi-robot localization is based on determining the relative poses of robots. We propose to compare the current perception data (laser scans) of one robot with the perception data acquired by another robot while moving. If the data match with high probability then the first robot is probably located on the path made by the second one.

The problem of matching perception data from different robots is a kind of revisiting problem. A robot should decide, whether it is at a location previously visited by another robot. To realize the mentioned scenario we introduce two functions $F : S \longrightarrow \mathfrak{P}(\mathfrak{S} \times S \times \mathbb{N})$ and $G : S \times L \longrightarrow [0,1]$, S, L and $\mathfrak{S}$ are the sets of all possible laser range scans poses and sample sets over poses respectively, $\mathbb{N}$ is the set of natural numbers and $\mathfrak{P}$ denotes the power set. Function G is defined as $G(s,\Delta l) = \int_{l'-\Delta l}^{l'+\Delta l} p(s|l,m)dl / \int p(s|l,m)dl$, where Δl is a small interval in the space of robot poses and $l' = \underset{l \in L}{argmax}\{p(s|l,m)\}$. The function returns the probability that a given scan s represents some limited area $[l'-\Delta l, l'+\Delta l]$ on the map (note that G is NOT a probability distribution). Intuitively G is a measure of "ambiguity" of a laser range scan s. Higher values of $G(s,\Delta l)$ indicate that s is less "ambiguous", that is s can probably be observed only within a small connected area in the map, while lower values mean that s is probably observed at several unconnected areas.

F is a kind of global repository where robots can store information on their own paths and retrieve that on the paths of other robots. Initially, $F(s) = \varnothing$ for all $s \in S$. As the robots start moving they estimate their local poses by means of a simplified version of the Rao-Blackwellised particle filter algorithm mentioned in Section II. The simplification here is that $\widehat{m}(l_{0:t-1}, s_{1:t-1}) = \widehat{m}(l_{t-1}, s_{t-1})$, i.e. we use only the most recent laser range scan as a map. If the j-th robot estimates a sample set S_t representing its local pose and reads a laser range scan s_t at time t then it sets $F(s_t) = F(s_t) \cup \{(S_t, s_t, j)\}$. In addition for all $(S', s', k) \in F(s_t)$ s.t. $k \neq j$ robot j first checks whether it should fuse its belief about the global pose with that of robot k (this could be done e.g. by comparing the entropies of both distributions) and whether the probability of a false positive path match is low, i.e. $1 - G(s_t, \Delta l) < a$, where Δl and a are some small (but not infinitely small) interval and threshold value respectively. The latter condition checks whether the local beliefs S_t and S' represent a same small area. This condition can be easily extended to longer pieces of paths. We call the latter condition "path matching of robot j with respect to robot k" for further reference.

We learn the functions F and G in a way similar to (Thrun et al., 2001). First, we randomly generate a sufficient number of poses in the map (one million in our current implementation) and then by means of ray-tracing and sampling from the perception model $p(s|l,m)$ compute "noisy" scans corresponding to the poses. Each scan is then transformed as mentioned above and a kd-tree is built over the transformed scan vectors. Initially, we store at each leaf of the tree the set of poses corresponding to the scan vectors making up the leaf. Each scan is then transformed into a lower dimensional space using the Haar wavelet transform (Jensen and la Cour-Harbo, 2001) and a kd-tree is built over the transformed scan vectors.

Our algorithm for the multi-robot localization is summarized below. It is easy to see, that the run-time complexity of the algorithm (after the mentioned functions has been learned) is not significantly higher than that of single robot approaches, because the path matching does not depend on the number of samples and is done very efficiently by looking up in a binary tree. The algorithm is robust in the sense that in case of communication failure the robots just do the single robot localization before they can communicate again, after which they simply continue running the multi-robot algorithm.

Multi-MCL

1. **for each** robot i **do**
2. estimate the current global pose sample set S_i by MCL;
3. estimate the current local pose sample set S'_i
   by the simplified Rao-Blackwellised particle filter algorithm;
4. $F(s_i) = F(s_i) \cup \{(S'_i, s_i, i)\}$,
   where s_i is the current laser range scan;
5. **for each** $(S'_j, s_j, j) \in F(s_i)$, $j \neq i$ **do**
6. **if** path matching of robot i with respect to robot j
   and j is localized better than i **do**
7. localize robot i in a small map defined by $\widehat{m}((0,0,0), s_j)$
   and compute sample set S_{ij} representing
   the relative pose of robot i with respect to robot j;
8. refine S_i by setting $S_i = S_i \cap S''_i$,
   where S''_i is a sample set resulting from translating the
   global pose samples of robot j by relative poses in S_{ij};

4. Experimental Results

We carried out a number of simulations to see how well the algorithm works in practice. We used the robot simulation software of the Aachen RoboCup team "AllemaniACs" (Ferrein et al., 2004). The software can produce noisy odometry and laser scan data and therefore can very accurately simulate real robots. We carried out several simulation runs for groups from 2 to 5 robots and for each case computed the average time per robot required to globally localize a robot both by the single-robot MCL and multi-robot MCL. The simulations showed that the average localization time per robot is approximately linearly reduced with an increased number of robots. This may be explained by the fact that we let the robots start at locations distributed quite uniformly in the map and because the robots could determine their relative poses already at earlier stages, they could use the considerable orthogonality in their perception data to exclude false hypotheses (samples) faster.

5. Conclusion

We presented a new Monte Carlo localization algorithm for multi-robot systems. Our approach differs from existing ones in the field in that it does not require the detection and identification of robots by each other for determining their relative poses, which are used to mutually refine the posterior distributions over global poses of the robots. The simulations showed an average reduction in the localization time linear in the number of robots.

References

Burgard, W., Cremers, A. B., Fox, D., Hähnel, D., Lakemeyer, G., Schulz, D., Steiner, W., and Thrun, S. (2000). Experiences with an interactive museum tour-guide robot. *Artificial Intelligence*, 114(1-2).

Doucet, A. (1998). On sequential simulation-based methods for bayesian filtering. Technical Report CUED/F-INFENGS/TR 310, Cambridge University, Department of Engineering, Cambridge, UK.

Doucet, A., de Freitas, J., Murphy, K., and Russel, S. (2000). Rao-blackwellised particle filtering for dynamic bayesian networks. *Proc. of the Conference on Uncertainty in Artificial Intelligence (UAI).*

Ferrein, A., Fritz, C., and Lakemeyer, G. (2004). Allemaniacs 2004 team description. http://robocup.rwth-aachen.de.

Fox, D., Burgard, W., Dellaert, F., and Thrun, S. (1999a). Monte carlo localization: Efficient position estimation for mobile robots. *Proc. of the Sixteenth National Conference on Artificial Intelligence (AAAI-99).*

Fox, D., Burgard, W., Kruppa, W., and Thrun, S. (2000). A probabilistic approach to collaborative multi-robot localization. *Autonomous Robotics, Special Issue on Heterogeneous Multi-Robot Systems*, 8(3):325–344.

Fox, D., Burgard, W., and Thrun, S. (1999b). Markov localization for mobile robots in dynamic environments. *Journal of Artificial Intelligence Research*, (11):391–427.

Fox, D., Ko, J., Stewart, B., and Konolige, K. (2003). The revisiting problem in mobile robot map building: A hierarchical bayesian approach. *In Proc. of the Conference on Uncertainty in Artificial Intelligence.*

Howard, A., Mataric, M., and Sukhatme, G. (2002). Localization for mobile robot teams: A distributed mle approach. *In Proc.of the 8-th International Symposium in Experimental Robotics (ISER'02).*

Jensen, A. and la Cour-Harbo, A. (2001). *Ripples in Mathematics: the Discrete Wavelet Transform*. Springer.

Kaelbling, L., Cassandra, A., and Kurien, J. (1996). Acting under uncertainty: Discrete bayesian models for mobile robot navigation. *In Proc. of the IEEE/RSJ Internatinal Conference on Intelligent Robots and Systems (IROS).*

Kurazume, R. and Hirose, S. (2000). An experimental study of a cooperative positioning system. *Autonomous Robots*, 8(1):43–52.

Montemerlo, M. and Thrun, S. (2003). Simultaneous localization and mapping with unknown data association using fastslam. *Proc.of the IEEE International Conference on Robotics and Automation (ICRA).*

Nourbaksh, I., Powers, R., and Birchfield, S. (1995). Dervish an office-navigating robot. *AI Magazine*, 16(2).

Roumeliotis, S. and Bekey, G. (2000). Collective localization: A distributed kalman filter approach. *In Proc. of the IEEE International Conference on Robotics and Automation*, 2:1800–1087.

Rubin, D. B. (1998). Using sir algorithm to simulate posterior distributions. In Bernanrdo, M., van De Groot, K., Lindley, D., and Smith, A., editors, *Bayesian Statistics 3*. Oxford University Press, Oxford, UK.

Simmons, R. and Koenig, S. (1997). Probabilistic robot navigation in partially observable environments. *In Proc. of the International Joint Conference on Artificial Intelligence (IJCAI).*

Thrun, S., Fox, D., Burgard, W., and Dellaert, F. (2001). Robust monte carlo localization for mobile robots. *Artificial Intelligence.*

A DIALOGUE-BASED APPROACH TO MULTI-ROBOT TEAM CONTROL

Nathanael Chambers, James Allen, Lucian Galescu, Hyuckchul Jung
Institute for Human and Machine Cognition
40 S. Alcaniz Street
Pensacola, FL 32502
{nchambers, jallen, lgalescu, hjung} @ihmc.us

Abstract This paper describes an approach to integrating a human into a team of robots through a dialogue-based planning assistant. We describe how the Rochester Interactive Planning System (TRIPS) is able to recognize and translate a user's intentions into collaborative problem solving acts. The user interacts naturally through language and avoids complex machine interfaces as TRIPS manages the appropriate lower level robotic commands and semantics. The system acts as a mediator to the user, managing all coordination and agent communication for a vast number of robots that would normally overwhelm a single user.

Keywords: agents, dialogue-based interaction, human-agent interface, mine search, multi-robot control, teamwork

1. Introduction

This paper describes recent work attempting to naturally integrate a human into a team of robotic agents. We describe the motivations for using a dialogue-based planning system to assist in robot communication and management. The system allows the user to interact with multiple robots through a single entity, thus reducing the burden of communicating simultaneously with multiple robots.

Although we describe a scenario of a single human and multiple robotic agents, much of this work may be extended to teams with varying numbers of humans and agents. Controlling a team of robots is inherently more difficult than single robot control: the human tries to manage multiple robots and to efficiently communicate with them, and the robots try to determine how to accomplish a global objective. As two or more robots attempt to communicate with the human, there is often cognitive overload for the human. A system that understands the human's goals and assists in building plans is essential in a framework that frees the user to work on higher level problems.

L.E. Parker et al. (eds.),
Multi-Robot Systems. From Swarms to Intelligent Automata. Volume III, 257–262.

TRIPS acts as a mediator who collaborates with the human user to construct plans that can achieve the user's goals, communicating with the robots only when necessary. Alerts, notifications, and queries coming from the robots must go through the mediator who then presents them to the user through dialogue and/or an appropriate graphical interface. The user is freed from monitoring each robot, performing low-level system commands, and grasping difficult semantics and communication languages that robots use. One of the most general models of this interaction is an intent-driven, human-centered model based on natural human-human dialogue.

An intent-driven approach to human-robot interaction is relatively new, having been applied mainly to single software agent planners. We are abstracting the general properties of such an approach and applying it to the new domain of controlling and interacting with multiple agents (who may have plans of their own).

2. Background

TRIPS is a mixed-initiative planning assistant that interacts with a single user through dialogue and supplemental graphical interfaces. Its central components are based on a domain independent representation of linguistic semantics, illocutionary acts, and a collaborative problem-solving model. The system is tailored to individual domains through a domain specific ontology mapper that maps the independent into the specific domain's semantics. The interaction between the core reasoning components drives the system in assisting the user in *planning* and *execution*. Planning with the user involves creating collaborative problem solving acts and maintaining the user's goals, working with him/her to accomplish their needs. Communication with robotic agents is accomplished through a backend module that translates the user's high level intentions into low-level commands that robots can understand. Figure 1 gives a brief description of the main modules. For a complete description of the TRIPS architecture, see (Allen et al., 2001, Allen et al., 2002).

Our approach uses the KAoS (Bradshaw et al., 2004) environment for robot communication and control. KAoS provides the physical communication framework and a set of policy services that allow the user to adjust the capabilities of agents by restricting their behavior, access to resources, and task initiative.

3. Mine Search Scenario

We have applied the TRIPS architecture to a mockup mine detection scenario (using unmanned ground vehicles), an area of great concern to the Navy. One solution to this difficult task is to deploy a (potentially large) number of robots to quickly search and identify mines, allowing humans to then neutralize them. The robots work with humans through mounted cameras to identify

Interpretation Manager (IM)	Coordinates the contextual interpretation of utterances.
Behavioral Agent (BA)	Maintains the collaborative problem solving state between the user, system, and robots.
Generation Manager (GM)	Realizes planned speech acts from the BA into spoken utterances and/or GUI events.
Discourse Context	Maintains discourse history to support reference and ellipsis. Identifies discourse acts, obligations, and expectations.
Ontology Manager (OM)	Maintains the domain independent logical form (LF) and the domain specific languages and provides a mapping between them.
SLIK	Simple Logical Interface to KAoS. Provides mappings between collaborative problem solving acts and external agent communications.

Figure 1. Key modules in TRIPS for interacting with a team of robots

unknown objects. This is an excellent example of a mixed-initiative task and requires coordination between robots and humans. The task also requires surface robots to facilitate in communication from the ship to underwater robots because of their extremely limited bandwidth. The robots need a significant level of autonomy to act in the absence of human intervention. A sample of our system interacting with the user can be seen in figure 2.

User:	Find a clear lane through this area <selects rectangular region on map>
TRIPS:	ok
TRIPS:	Alex Richard and Robert are available. <highlights robots on the map> Which do you want to use?
User:	all three

Figure 2. A user's interaction with TRIPS, initiating a team of robots to search an area for mines.

4. Advantages of a Dialogue-Based Approach

We believe a model for an effective human-robot collaboration must include a planning mediator that is able to interpret a dialogue-based interaction while

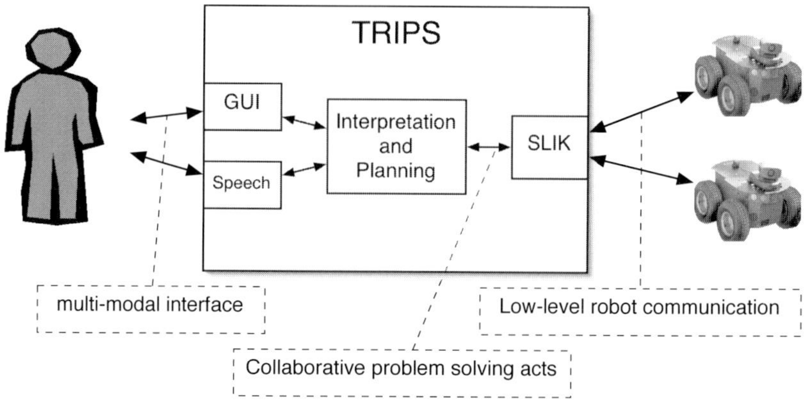

Figure 3. Control flow between the user, TRIPS, and a team of robots.

hiding the user from lower-level robotic requirements. Figure 3 provides a description of our model.

4.1 Dialogue-Based Multi-Modal

Experience with this system has shown that the user greatly benefits from working with a single dialogue-based collaborative agent. As teams of agents grow larger, the user has difficulty understanding who he is/should be talking to. An assisting and planning collaborator reduces the complexity and allows the user to place the burden on TRIPS. Direct human to agent communication can be done through speaking agent names (we used human names in this scenario) or by selecting them on a GUI. In addition, the user can direct large numbers of robots at once and speak of them as a group or a single entity. The multi-modal aspect of TRIPS allows the user to speak in terms that are natural to humans, and use GUIs for deictic reference.

Speak	**GUI action**
'have this robot search a new leg'	<click robot icon on map>
'use these robots'	<highlight multiple robots on map>

4.2 A Planning Mediator

A significant problem in dealing with large teams of agents (or even mixed humans and agents) is that the agents don't necessarily act as helpful collaborators. They may perform complementary tasks with the human to solve a certain problem, but they don't help the user formulate his own plan and method of performing a task. The dialogue-based planning assistant serves as a helpful interface to robots, but it can also build and maintain plans collaboratively with the user. This kind of assistant is not usually found in the other agents of a team.

Secondly, the mine detection scenario makes it clear that the role of TRIPS as a mediator in limiting how information is conveyed to the user is essential to a manageable interface. While many research groups have implemented proxies, it is important to create a mediator that not only has a representation of the user's desires, but also of the user's problem-solving state in order to effectively present and hide varying levels of information coming from the external world.

4.3 Robot Communication

Robotic agents are often relatively primitive in their communication abilities. They may be very sophisticated in regard to the task at hand, but the communication barrier is a difficult one to overcome and is usually less of a priority to robot developers. For these reasons, we use the **Simple Logical Interface to KAoS** (SLIK) to act as a gateway to KAoS, keeping the undesired low level communication hidden from both TRIPS and the human user. SLIK receives a set of collaborative problem-solving acts from the Behavioral Agent in TRIPS and typically translates a *query* into an information *command* that is sent to a robot (our robots, like most robots, are command driven and currently understand OWL). The BA can send queries and receive answers from SLIK as if the agents themselves understood them. Likewise, TRIPS receives input from the robots as if they also communicate in higher level acts. This approach of a single gateway is beneficial in that it can handle requests and commands to teams of robots, not just a single one, in a high level semantic language.

5. Related Work

Most work involving one human and multiple robots use the human in a supervisory role with an emphasis on GUIs (Blake et al., 2000, Jones et al., 2002, Payne et al., 2000). Our approach differs in that the GUI modality is used to supplement the interaction, not dominate it.

Several groups have developed *proxies* (Kortenkamp et al., 2002, Scerri et al., 2003, Martin et al., 2003) for the human that help filter information for the user by maintaining some sort of user model. The work by Martin et al. is perhaps the most similar in that they use a global system planner with pre-defined plans for all the *agents*. Our approach contains a personal planner driven by a model of the *user's* intentions.

6. Conclusion

We have described a general approach for a dialogue-based mediator to assist humans in communicating with non-human-centric robots. By understanding the user's intentions and the current problem solving state, this assistant can reason about the status of multiple agents and lessen the cognitive load on the user. We believe multi-modal mixed-initiative planning assistants like ours are needed to achieve effective collaboration between a mixed team of humans and agents.

Acknowledgments

We thank the KAoS group at IHMC for their support and helpful insight into human-agent collaboration. This work was supported in part by ONR grant N000140310780.

References

Allen, J., Blaylock, N., and Ferguson, G. (2002). A problem solving model for collaborative agents. In *Proceedings of AAMAS-02*, Bologna, Italy.

Allen, J., Ferguson, G., and Stent, A. (2001). An architecture for more realistic conversational systems. In *Intelligent User Interfaces*, pages 1–8.

Blake, M. A., Sorensen, G. A., Archibald, J. K., and Beard, R. W. (2000). Human-assisted capture-the-flag in an urban environment. In *IEEE Int. Conf. on Robotics and Automation*.

Bradshaw, J. M., Jung, H., Kulkarni, S., Allen, J., Bunch, L., Chambers, N., Feltovich, P., Galescu, L., Jeffers, R., Johnson, M., Taysom, W., and Uszok, A. (2004). Toward trustworthy adjustable autonomy and mixed-initiative interaction in kaos. In *Proceedings of the AAMAS Workshop on Trust in Agent Societies*.

Jones, H. L., Rock, S. M., Burns, D., and Morris, S. (2002). Autonomous robots in swat applications: Research, design, and operations challenges. In *Proceedings of the 2002 Symposium for AUVSI*, Orlando, USA.

Kortenkamp, D., Schreckenghost, D., and Martin, C. (2002). User interaction with multi-robot systems. In *Workshop on Multi-Robot Systems*.

Martin, C., Schreckenghost, D., Bonasso, P., Kortenkamp, D., Milam, T., and Thronesbery, C. (2003). Aiding collaboration among humans and complex software agents. In *AAAI Spring Symposium on Human Interaction with Autonomous Systems in Complex Environments*.

Payne, T., Sycara, K., Lewis, M., Lenox, T., and Hahn, S. (2000). Varying the user interaction within multi-agent systems. *Autonomous Agents*.

Scerri, P., Pynadath, D. V., Johnson, L., Rosenbloom, P., Schurr, N., Si, M., and Tambe, M. (2003). A prototype infrastructure for distributed robot-agent-person teams. In *Proceedings of AAMAS-03*.

HYBRID FREE-SPACE OPTICS/RADIO FREQUENCY (FSO/RF) NETWORKS FOR MOBILE ROBOT TEAMS

Jason Derenick, Christopher Thorne, and John Spletzer

Lehigh University
Bethlehem, PA USA
spletzer@cse.lehigh.edu

Abstract In this paper, we introduce a *hybrid free-space optics/radio frequency (FSO/RF)* networking paradigm for mobile robot teams. We believe that such a model will emerge as a consequence of inherent limitations of RF based approaches. FSO technology has the potential to provide tremendous increases in per-node throughput for a mobile ad-hoc network (MANET). To motivate this paradigm, we first provide a brief background on FSO and discuss potential applications where its capabilities could be well leveraged. We then provide initial experimental results from autonomous deployments of a hybrid FSO/RF MANET with real-time video data routed across both optical and RF links.

Keywords: Free-space optics, FSO, MANET

1. Introduction

Over the past decade, mobile robotics research has yielded significant improvements in sensing and control algorithms, enabling levels of autonomy and robustness never before realized ((Fox et al., 2000), (Dissanayake et al., 2001)). These advances were driven in part by the maturation of the underlying technologies: the unimpeded progression of Moore's law, improvements in sensor and actuator technology, and the advent of wireless network communication.

Additional improvements in embedded devices and micro MEMS technology have enabled a second related technology: *wireless sensor networks* (WSN). In a WSN, each sensor node is capable of sensing its environment, and collecting/processing data ((Nemeroff et al., 2001)). Ideally, the WSN is fully connected. Each node has a minimum connectivity degree of three or four, allowing for direct positions estimates. Node connectivity is associative, and bandwidth constraints are insignificant compared to available computational

L.E. Parker et al. (eds.),
Multi-Robot Systems. From Swarms to Intelligent Automata. Volume III, 263–268.

resources. When such idealisms fail to hold, the WSN is congested, disconnected, and incapable of localizing itself in the environment.

The ability to alleviate many weaknesses in static WSNs by introducing mobility has lead to the fusion of mobile robotics and WSN technologies, and the formation of *Mobile Ad-hoc Networks* (MANETs). Mobile WSN can achieve otherwise unobtainable levels of robustness and performance ((Grossglauer and Tse, 2001)). However, MANETs are still bound by the provable limits in per-node throughput for radio frequency (RF) based communications ((Gupta and Kumar, 2000)). We predict that these limitations will lead to the emergence of *hybrid free-space optics/radio frequency (FSO/RF) MANETs*. While both technologies possess significant limitations, FSO and RF are complementary and have the potential to mitigate each other's weaknesses.

2. Background and Related Work

2.1 Free Space Optics (FSO): A Brief Overview

FSO is a commercial technology used primarily in static configurations for high bandwidth, wireless communication links ((Willebrand and Ghuman, 2002)). Degradations in FSO link performance from extreme atmospheric conditions (*e.g.*, heavy fog, smoke, *etc.*) lead to the development of hybrid FSO/RF networks. This was motivated by the need to preserve carrier class link availability (99.999%). In this paradigm, the RF link serves as a low-bandwidth backup to the primary optical link

In FSO, an optical transceiver is placed on each side of a transmission path to establish the network link. The transmitter is typically an infrared (IR) laser or LED which emits a modulated IR signal. This is captured by the receiver at the remote link side. FSO links can be full duplex, meaning that they can transmit and receive data simultaneously.

Several transmitter/receiver link configurations are possible: directed, diffuse, and hybrid (see Figure 1. The former offer the greatest link ranges, and are used for "last mile" commercial network links ((Barry, 1994)). Diffuse models are often used in "secure" wireless LANs, as unlike their RF counterparts communications can be readily be constrained to a single room. However, diffusing the transmitted signal significantly reduces transmission ranges (to tens of meters).

2.2 Motivation

FSO offers several advantages over RF based technologies. The most significant of these is extremely high throughput across long link distances. Widespread RF technologies (*e.g.*, 802.11x) are limited to link throughputs on the order of 10s of Mbps. Even much anticipated ultra wideband (UWB) technol-

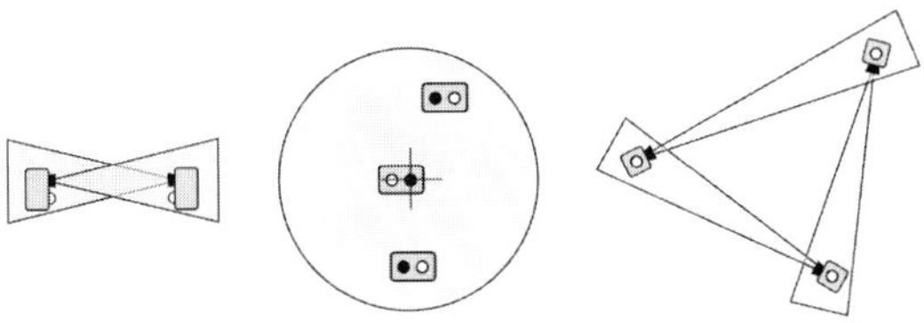

Figure 1. FSO link geometries: directed transmitter/receiver pair (left), diffuse transmitter/receivers (center), and hybrid directed transmitter with diffuse receiver (right).

ogy - with theoretical throughput of 100s Mbps - drops to levels lower than 802.11a at modest ranges ($r \geq 15m$) ((Wilson, 2002)). In contrast, FSO offers throughputs of several Gbps in currently fielded systems with link distances of a kilometer or more. Fig. 2 highlights the range vs. throughput characteristics of these three mediums. It should be noted that the achievable FSO link distance will be a function of atmospheric conditions. Those reported here are based upon a signal attenuation of 17 dB/km. This corresponds to extremely heavy rainfall (10 cm/hour) or light fog conditions.

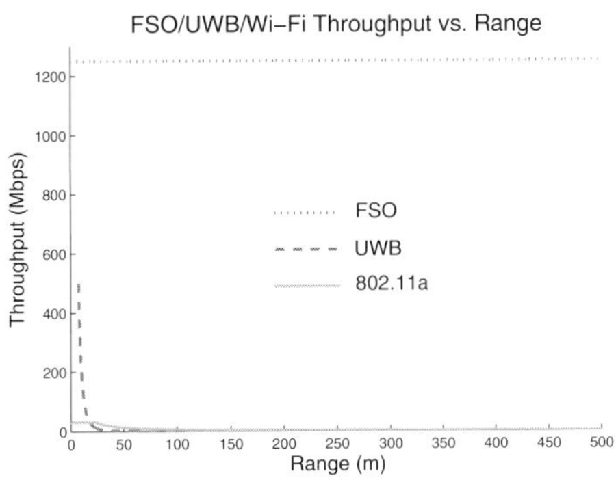

Figure 2. Throughput vs. range for FSO, UWB, and 802.11a technologies. Commercially available FSO transceivers can maintain gigabit+ throughput at distances of 1 km or greater. (Sources: Intel R&D and Canon Canonbeam DT-130).

However, in the realm of mobile networking it has lost in the mainstream computing market to RF technology. This can be attributed to one reason: the requirement for optical links to maintain line-of-sight (LOS). As such, FSO cannot replace RF wireless communications. Instead, the two technologies will serve in complementary modes in a hybrid FSO/RF network. For example, a MANET so equipped could prove invaluable for disaster relief - when some natural or man-made event (*e.g.*, an earthquake) destroys critical infrastructure. In such an application, a team of robots could automatically deploy and provide carrier grade patches to a Metropolitan Area Network (MAN). This could even be accomplished under conditions unsafe for human operators.

Other potential advantages of FSO enabled MANETs include improved node localization. An FSO link constrains the positions of link ends through relative *bearings*. These provide significantly stronger constraints on node positions than traditional range measurements alone ((Chintalapudi et al., 2004)). FSO also offers low per-bit transmission cost. The directed nature makes it inherently secure, the transmission provides no RF signature, and is also immune to jamming and electromagnetic interference ((Barry, 1994)).

3. Initial Experimental Results with a FSO/RF MANET

To demonstrate the feasibility of approach, a series of experiments was conducted using "Gollum" and "Balrog" - our in-house mobile robot research platforms (Fig. 3). Each was equipped with a *LaserWire* 100 Mbps optical head, alongside a traditional 802.11 based transceiver.

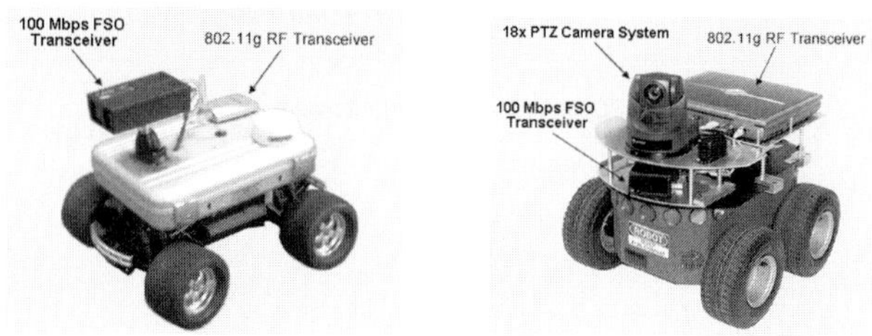

Figure 3. Mobile research platforms equipped with 100 Mbps FSO and 802.11g RF transceivers. "Balrog" (right) also employs an 18x pan-tilt-zoom (PTZ) system for locating link partners at ranges up to 100 meters.

In our preliminary experiments, the objective was to deploy and instantiate a FSO/RF network. The operational scenario is illustrated in Fig. 4, the premise of which is a remote surveillance task. In the initial configuration, the two mobile platforms were co-located at the far west (left) side of Packard Laboratory. Upon orders from the Command Center (CC), Balrog autonomously deployed to the far east (right) side of the building, traversing a distance of approximately 45 meters. It's task was to transmit local video data from an area of interest (the surveillance zone) back to the CC. The subsequent separation distances between the nodes was such that: 1) No RF link existed between Balrog and the CC, 2) The RF link between Gollum and Balrog provided insufficient throughput for the video stream. These facts necessitated the establishment of an optical link between the two mobile nodes. Thus, in addition to acting as a physical link between the RF and FSO networks, Gollum also had to be capable of forwarding packets between the networks over the correct network interface. This is a routing domain problem that was rectified by configuring Gollum with a packet forwarding (routing) table. A total of 10 deployments

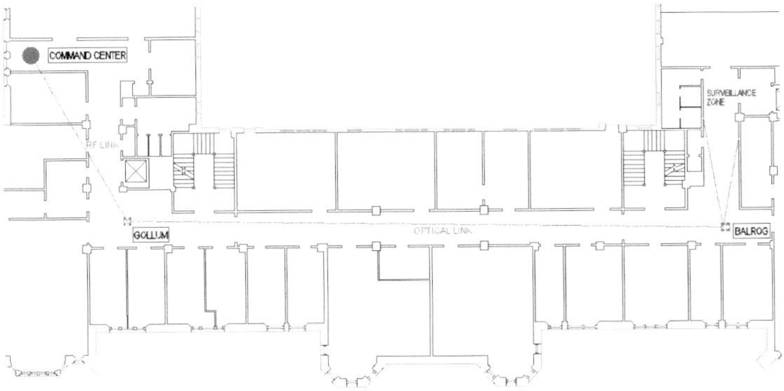

Figure 4. Deployment of a FSO/RF MANET. On orders from the Command Center (CC) (upper left), Balrog deploys to west (right) side of Packard Lab. After establishing the FSO link with Gollum, surveillance video is routed back to the CC over both FSO and RF links.

were conducted. In each, the robots were successful in deploying and instantiating the hybrid network, and routing surveillance video back to the command center. A sample trial is illustrated in Figure 5. More detailed project information can be obtained at *http://www.cse.lehigh.edu/ jcd6/fso/*.

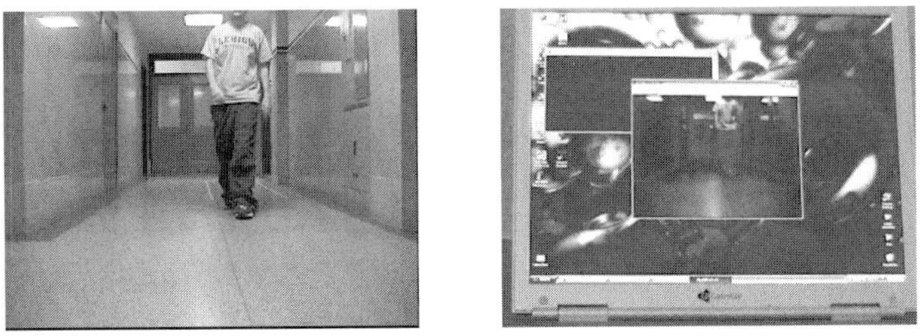

Figure 5. Local surveillance video captured by Balrog (left) and the remote real-time video as viewed at the Command Center

4. Conclusions and Future Work

In this paper, we introduced a hybrid FSO/RF network paradigm for mobile robot teams. Our primary motivation is the tremendous throughput provided by FSO over long link distances. However, the technology also offers the potential for improved localization performance, low power consumption, and secure/robust transmissions.

To demonstrate the feasibility of this model, preliminary experiments were conducted whereby a FSO/RF MANET was deployed and FSO/RF links established dynamically. Our future work includes the development of a hierarchical vision/FSO based link acquisition system (LAS), and adaptive beam divergence to support mobile operations

References

Barry, J. (1994). *Wireless Infrared Communications*. Kluwer Academic Publishers.

Chintalapudi, K., Dhariwal, A., Govindan, R., and Sukhatme, G. (2004). Ad-hoc localization using ranging and sectoring. In *IEEE INFOCOM*.

Dissanayake, G., Newman, P., Durrant-Whyte, H., and Csorba, M. (2001). A solution to the simultaneous localization and map building. *IEEE Transactions on Robotics and Autonomation*, 17(3):229–241.

Fox, D., Burgard, W., Kruppa, H., and Thrun, S. (2000). A probablistic approach to collaborative multi-robot localization. *Autonomous Robots: Special Issue on Heterogeneous Multi-Robot Systems*, 8(3):325–344.

Grossglauer, M. and Tse, D. (2001). Mobility increases the capacity of ad-hoc wireless networks. In *IEEE INFOCOM*.

Gupta, P. and Kumar, P. (2000). The capacity of wireless networks. *IEEE Transactions on Information Theory*.

Nemeroff, J., Garcia, L., Hampel, D., and DiPierro, S. (2001). Application of sensor network communications. In *IEEE MILCOM*.

Willebrand, H. and Ghuman, B. (2002). *Free Space Optics: Enabling Optical Connectivity in Today's Networks*. Sams Publishing.

Wilson, J. (2002). Ultra-wideband / a disruptive rf technology? Technical report, Intel Research and Development.

SWARMING UAVS BEHAVIOR HIERARCHY

Kuo-Chi Lin
University of Central Florida
3280 Progress Drive
Orlando, FL 32826, U.S.A
klin@pegasus.cc.ucf.edu

Abstract: This paper uses a behavioral hierarchy approach to reduce the mission solution space and make the mission design easier. A UAV behavioral hierarchy is suggested. A collection of lower level swarming behaviors can be designed under this hierarchy. Mission design can be simplified by picking the right combination of those swarming behaviors.

Keywords: Swarming, UAV, Multiple Agents

1. Introduction

The use of Unmanned Aerial Vehicles (UAVs) in the battlefield has gained more and more attentions. The current operation takes a team of human operators to control one UAV remotely. This approach becomes impractical if a large number of UAVs is used in the same battlefield. Not only more human operators are needed, but also the collaboration among human teams is a difficult issue. Another problem to consider is the cost. To deploy a group of very intelligent and multi-functional UAVs can be very expensive.

What are the alternatives? Besides the multi-functional, fully autonomous, highly intelligent (and therefore expensive) UAVs, at the other end of the spectrum, the single-function, limited-intelligence low-cost ones are also surprisingly useful. This idea is inspired by the social insects such as ants (Bonabeau, E., et al, 1999, Lin, K. C., 2001). One ant, by itself, is powerless; but hundreds of them working together can accomplish difficult tasks. Some advantages of using a swarm of low-end UAVs are obvious: they are cheaper and easier to build. Another more important feature is that

L.E. Parker et al. (eds.),
Multi-Robot Systems. From Swarms to Intelligent Automata. Volume III, 269–275.

a mission carried out by them is more robust since the loss of a few robots due to malfunctions or damages (from enemies) may not jeopardize the mission. However, because of the limited capability of the low-end robots, how to make them work together remains a difficult challenge. After all, scientists have not fully understood how ants work as a large team.

The author suggests an approach that uses the combination of the lower level behaviors to achieve the higher level objectives. To use this approach, the first thing needed is a hierarchy of the swarming UAVs behaviors. The author suggests the following hierarchy:

- High-level behaviors (e.g., strategic maneuvers, resources distribution)
- Tactical-level behaviors (e.g., reconnaissance, surveillance (Lin, K. C., 2001), suppression)
- Swarming
- Individual behaviors (e.g., avoidance, tracking, homing, following)

This paper will focus on the swarming behaviors.

2. Swarming UAV Model

The definition of "swarm", according to Clough (Clough, B., 2002), is

> "A collection of autonomous individuals relying on local sensing and reactive behaviors interacting such that a global behavior emerges from the interactions".

This definition makes distinguishes between a swarm and a team. Teams are deliberate behaviors – each member has a role to accomplish, knows what that is, knows what the other member's roles are, and knows how they relate as the task is accomplished. They have a plan. For a swarm, however, the global behaviors emerge from the collection of individual behaviors, which are local and reactive.

Figure 1 compares these two concepts. A team of nine UAVs start with a diamond formation. When they encounter an obstacle, each member maneuvers around it. Afterwards, the team returns to its original formation (Figure 1(a)). But for a swarm, each member only tries to stay close as a swarm (Figure 1(b)). Those two examples may be simplistic, but can show the idea.

The advantages of using a swarm over a team are

- Robustness. The loss of a few UAVs due to malfunctions or damages by the enemies may not jeopardize the mission;
- Cost-effectiveness. A swarm of "dumb" UAVs can do more things than one "very smart" UAV yet cost less;
- Scalability. The missions are easier to scale up or down.

From the definition, the UAV swarm is modeled as:

- The UAV swarm is homogeneous except for a few specialists, if deeded;
- Each UAV only responds to local situations or threats based on the sensory inputs.
- The UAVs are controlled by a set of behavioral rules.
- Human controllers, either centralized or distributed, intervene only when necessary.

In the model, each UAV is reactive according to the behavioral rules. The question is how to design the rules so that these local reactive motions can emerge the global behaviors of the swarm that can accomplish the mission. Because of the complexity of the UAV interactions in the swarm, the solution space may be too large to search.

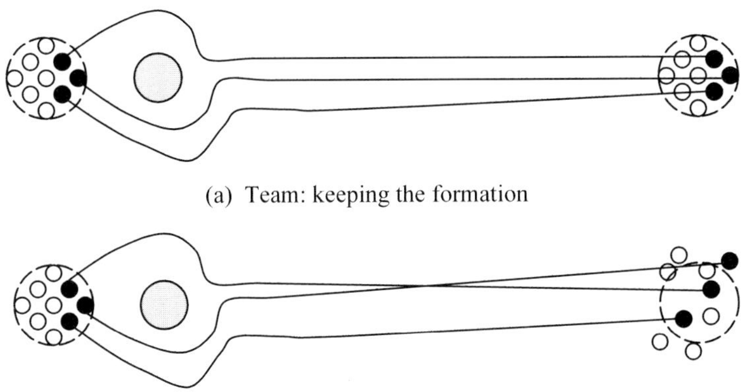

(a) Team: keeping the formation

(b) Swarm: staying close but no formation

Figure 1. Comparision between a team and a swarm.

3. Mission Design

The approach used in the paper is based on the following propositions. If each UAV's low-level behaviors are properly designed, the swarm can exhibit proper collective low-level behaviors. The higher-level, for example, the tactical-level, behaviors of the swarm can be the proper combination of sequences of low-level behaviors.

Based on the propositions, the design procedure is given by:

- Choose the higher-level behaviors needed for the mission;
- Combine the necessary low-level behaviors to form those higher-level behaviors;

- Design the controls of individual UAV to have the proper low-level behaviors;
- Close the loop for the optimization.

It can be seen from this procedure, the solution space is narrowed down to the individual UAV's low-level behaviors.

4. Swarming Behavioral Hierarchy

The author suggests a behavioral hierarchy as shown in Figure 2. In the boxes, the upper parts are the names of the behaviors and the lower parts are the individual behaviors which are common to this behavior and the levels below it. In other words, the behaviors in the lower level inherit the common individual behaviors from their ancestors. Each behavior is represented by its own name and its ancestors, separated by "dashes". For example, the behavior with a thicker box in Figure 2 is "Homing-Grouping-Swarming". To exhibit this behavior, all UAVs must have *at least* three individual behaviors, namely, Collision_Avoidance, Stay_Close, and Target_Track.

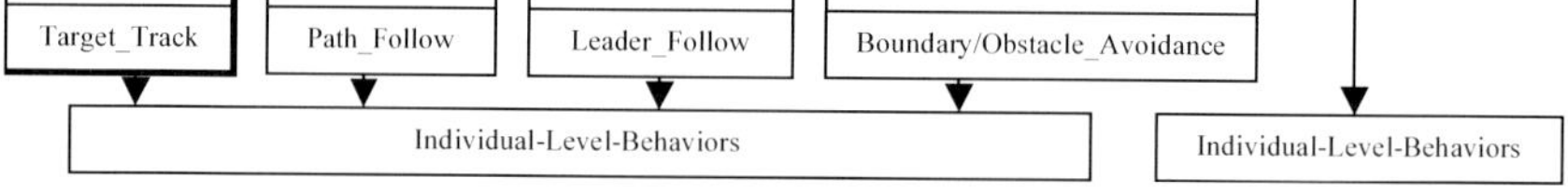

Figure 2. UAV swarming behaviors hierarchy.

To substantiate those collective behaviors, each UAV is controlled by a set of behavioral rules, such as Collision_Avoidance, Stay_Close, and Target_Track in the above example. Each rule is assigned a priority. A high priority rule overwrites the lower priority rules. By assigning priorities differently, the collective behaviors will be different. Also, there are parameters associated with the rules. For example, the Stay_Close rule has a radius associate with it. Therefore, each behavior can have many substantiated behaviors. In the mission design stage, the optimal combinations of behaviors with the parameters associated with them are chosen.

5. Example Behaviors

The behaviors of Wondering-Grouping-Swarming are used as examples. The scenario is when the swarm is approaching a boundary. Figure 3(a) shows the simulation result of the swarming behavior #1: each UAV has three individual behaviors with priorities from high to low: Collision_Avoidance, Boundary_Avoidance, and Stay_Close. The broken line represents the line that the UAVs detect the boundary, which is represented by the solid line. As shown in the figure, some UAVs go out of boundary to avoid other UAVs. If staying inside the boundary is very important, the Boundary_Avoidance individual behavior can be assigned the highest priority. Figure 3(b) shows the simulation result (behavior #2). Most UAVs have stayed inbound all the time. The tradeoff is that the probability of collisions among UAVs may be higher.

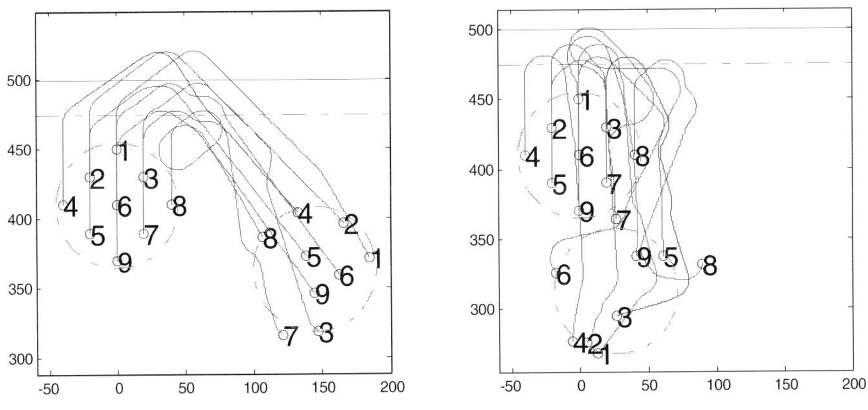

(a) (b)

Figure 3. Wondering-Grouping-Swarming behaviors.

6. Example Mission

Figure 4 shows an example mission. A swarm of UAVs leave from the left-side starting point to survey the rectangular area on the right, with an area to avoid and a boundary to stay within. When the swarm first leaves the starting point, Homing-Group-Swarm is used to move toward the target area. When the area to avoid is detected, Wondering-Group-Swarming with emphasis on obstacle_avoidance is used to avoid the area. Right after that, the upper boundary is detected; Wondering-Grouping-Swarming with emphasis on boundary_avoidance is used. After turning back from the boundary, Homing-Grouping-Swarming is used to move toward the target area. After entering the area to survey, Disperse-Swarming is used to spread the UAVs out and survey the area. In this behavior, each UAV has the individual behaviors of Obstacle_Avoidance and Boundary_Avoidance to stay in the area to survey.

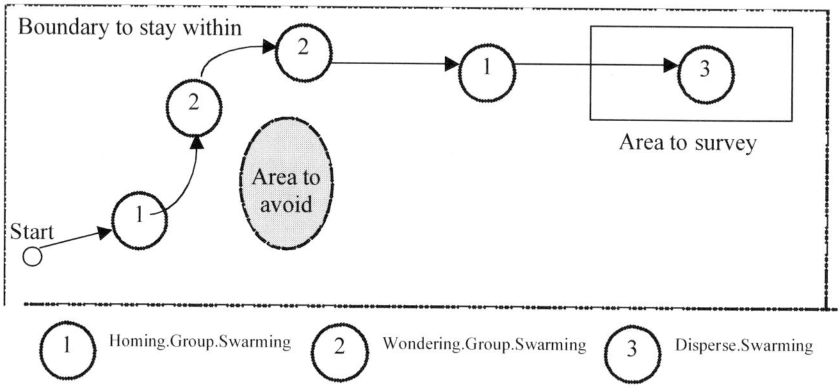

Figure 4. Surveillance mission.

7. Conclusion

This research has demonstrated that using the behavioral hierarchy, the solution space can be reduced to make the mission design easier. A collection of lower level swarming behaviors can be designed under this hierarchy. Each behavior can have a number of variable parameters associated with it. Mission design can be simplified by picking the right combination of those swarming behaviors with the proper parameters.

Acknowledgements

This research is partially sponsored by National Science Foundation and Air Force Research Laboratory.

References

Bonabeau, E., et al, (1999) "Swarm Intelligence: from natural to artificial systems", Oxford University Press, 1999.

Clough, B., (2002) "UAV Swarming? So What are Those Swarms, What are the Implications, and How Do We Handle Them?" Proceedings of the AUVSI Unmanned Systems Symposium, July 2002, Orlando, FL.

Lin, K. C., (2001) "Controlling a Swarm of UCAVs~A Genetic Algorithm Approach", Final Report for VFRP, Information Directorate, AFRL, 2001.

THE GNATS – LOW-COST EMBEDDED NETWORKS FOR SUPPORTING MOBILE ROBOTS

Keith J. O'Hara, Daniel B. Walker, and Tucker R. Balch
The BORG Lab
Georgia Institute of Technology
Atlanta, GA
{kjohara, danielbw, tucker} @cc.gatech.edu

Abstract We provide an overview of the GNATs project. This project is aimed at using tens to thousands of inexpensive networked devices embedded in the environment to support mobile robot applications. We provide motivation for building these types of systems, introduce a development platform we have developed, review some of our and others' previous work on using embedded networks to support robots, and outline directions for this line of research.

Keywords: Pervasive Computing, Sensor Networks, Multi-Robot Systems

1. Introduction

Pervasive networks of computing, communicating, and sensing devices will be embedded in future environments. These devices will include the likes of RFIDs, active badges, and sensor networks. For the most part, these devices are framed in the context of enabling and supporting human activities. We posit that these networks can also support robot systems, and particularly, mobile robot systems. In fact, we believe these networks will be so useful for mobile robots, that even when this infrastructure is not already available (e.g. space exploration) robots should expend the resources to deploy them as an early part of the mission.

Embedded networks can aid robots in completing their tasks, primarily by providing communication and coordination services, and possibly computation and sensing services. We feel this heterogeneous system of embedded devices and mobile robots puts a natural constraint on the design space of multi-robot systems. The embedded network serves as a pervasive communication, computation, and coordination fabric, while the mobile robots provide sensing and actuation.

L.E. Parker et al. (eds.),
Multi-Robot Systems. From Swarms to Intelligent Automata. Volume III, 277–282.

Additionally, not only can pervasive networks support mobile robots, they can also be supported by mobile robots. The tedious tasks of deployment and maintenance of a thousand node network is a perfect application of autonomous robot technology.

One possible criticism of using embedded networks to support mobile robots is that of "engineering the environment". Roboticists have worked tirelessly to make robots truly autonomous, often meaning the robots act intelligently in unknown and unpredictable environments. By creating infrastructure to support mobile robots, it may seem as though we are sidestepping this aspect of autonomy. We believe that almost all natural autonomous creatures build and use artifacts to support them in their daily tasks. As examples, ants lay pheromone trails and humans create traffic light systems. We feel that mobile robots can do the same. And if we must use the term "engineer the environment" – rather than the roboticist engineering the environment, we do believe it is useful for the robots to engineer the environment. The robots and the embedded network should have a symbiotic relationship by supporting each other, often in an autonomous manner.

In previous simulation work we investigated the use of embedded networks to facilitate mobile robot activities (O'Hara and Balch, 2004b). We have implemented a hardware platform to realize these types of applications. The platform, the GNATs[1], are low cost devices, allowing us to build a large number of them, and are highly configurable. The GNATs are intended to be used as a massively parallel system for computation, communication, and coordination in supporting mobile robots. The simplicity of the GNATs due to their specialization for mobile robot applications allows us to build them for a price an order of magnitude less than the Motes. This allows us to experiment with very large-scale systems.

2. The Hardware Platform

We have implemented a hardware platform, called the GNATs, for building embedded networks to support mobile robots. The hardware design choices were made explicitly to enable them to support mobile robots. The GNATs consist of four infrared (IR) emitters, four IR receivers, two visible light LEDs, a button, a Microchip PIC16F87 microcontroller, and a 3V battery. The platform is pictured in Figure 1. The simplicity of the platform makes it very inexpensive, allowing us to build, and experiment with, a large number of devices.

Using infrared as the communication medium has multiple advantages and some disadvantages. Infrared is short-range and line-of-sight, these characteristics make is useful for storing environmentally sensitive information, often the most useful to mobile robots. Because environmental information is often

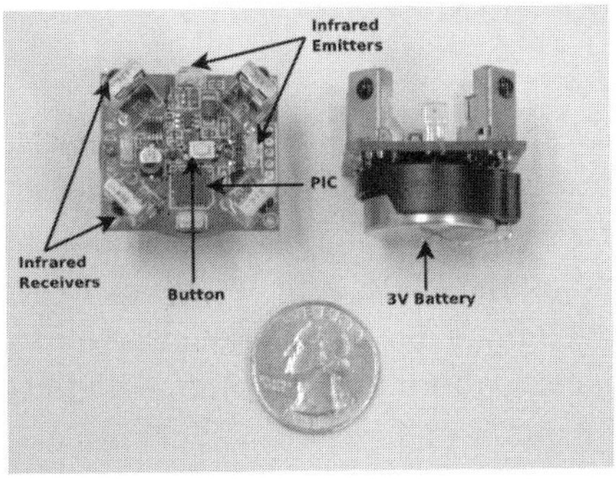

Figure 1. Two GNATs

local, we need a communication medium that respects this and keeps the information in context. This was the idea behind using infrared communication for "World-Embedded Computation" (Payton et al., 2001). Also, infrared is less power hungry than radio.

One disadvantage of infrared, as compared to radio, is its sensitivity to ambient, interfering, light sources. Many fluorescent lights (like the variety in our lab!) radiate infrared light that interferes with the infrared communication. Another disadvantage of infrared, as compared to radio, is its low data-rate. In general this is an disadvantage of infrared, but we don't feel this really impacts our applications since we don't imagine the network needing very high data-rates.

The GNATs can dynamically change their program code, their processor frequency, their communication output power and directionality, and turn off large parts of their circuitry when not in use for power-saving purposes. The GNATs also have a variety of sleep modes resulting in very long lifetimes. During these sleep-modes the GNATs can be configured to wake-up on timer or input (infrared activity, button) events.

Each device is less than $30 to build, enabling large-scale experimentation. The emitters' output power can be controlled with software allowing communication ranges of 1-5 meters. Also, the emitters can be individually addressed when sending messages, allowing the device to send messages in any combination of directions. Finally, the devices can write to their program memory permitting us to change the software on the devices on the fly, by a PC, or possibly a robot or other GNATs. The programming port can also be used for RS-232 serial communication. Using serial communication, one of the GNATs can be used as a communication device for a mobile robot. The mobile robot can carry a GNAT onboard to interact with other GNATs embedded throughout the environment.

3. Supporting Mobile Robots

Although, not explicitly directed at embedded networks, Parunak developed a technique for coordinating multiple unmanned air vehicles (UAVs) using synthetic pheromones and a multi-agent system (Parunak et al., 2002a, Parunak et al., 2002b). Inspired by pheromone communication in insects, they create potential fields for guiding the UAVs around threats to goal locations in a distributed manner. The technique they developed used uniformly placed (tiled as hexagons) "place" agents to store the pheromone and evaporate it over time, and "walker" agents to spread and react to the pheromone. The walker agents consisted of the UAV agents which physically move over the place agents and "ghost" agents which walk over the place agents virtually. The "place" agents could be implemented in the real world by using some kind of embedded network.

Several robotics researchers have proposed using embedded networks to support mobile robot applications. Both Batalin and Sukhatme (Batalin et al., 2004) and Li et al.(Li et al., 2003) have developed approaches to navigation using heterogeneous teams composed of mobile nodes and an embedded network. The network of embedded nodes, creates a "Navigation field" (Batalin and Sukhatme, 2003b), which mobile nodes can use to find the their way around. They differ in how they compute this navigation field. Batalin and Sukhatme use Distributed Value Iteration (Batalin and Sukhatme, 2003b). In their approach, the embedded nodes use estimated transition probabilities between nodes to compute the best direction to suggest to a mobile robot for moving between a start and goal node. These transition probabilities are established during deployment and both the robots and sensor nodes have synchronized direction sensors (e.g. digital compass). In addition to navigation, Batalin and Sukhatme have applied their technique to the multi-robot task allocation problem (Batalin and Sukhatme, 2003b).

Li et al. are able to generate an artificial potential field for navigation based on the obstacles and goals sensed by the network (Li et al., 2003). This potential field is guaranteed to deliver the mobile node to the goal location via an danger-free (obstacle-free) path. The field is created by the embedded nodes propagating goal-ness or danger to neighboring nodes. Both Batalin and Li used the Motes hardware platform for their physical experimentation.

In previous simulation studies we showed an embedded network supported effective cooperative multi-robot foraging by coordinating coverage patterns and by providing nearly optimal path planning without the network nodes having global knowledge or localization capabilities (O'Hara and Balch, 2004b). The embedded network created navigation networks for guiding mobile robots in various tasks such as coverage, recruitment, and path planning. Quantitative results illustrated the sensitivity of the approach to different network sizes, environmental complexities, and deployment configuration.

In addition, in previous work we developed and analyzed two different techniques for distributed path planning when the environment is dynamic (O'Hara and Balch, 2004a). One used global monitoring and the other focused communication. Both techniques were able to repair the plan when the environment changed and provided paths for a mobile robot to reach a goal. The first technique was able to respond to changes in the environment very quickly but did this at high communication cost. The second approach was able to respond to changes in the environment at the same speed, but with far fewer messages because it concentrated the messages along the path on which the robot currently resided.

A network of embedded nodes can also aid robots in coverage. Koenig (Koenig et al., 2001) and Wagner (Wagner et al., 1999) devise methods for doing parallel coverage using simple ant robots that communicate indirectly by leaving indicators in the environment. An embedded device can be used as this type of inexpensive indicator with the added advantage that they can communicate with each other. Batalin also uses communication nodes as "markers" in aiding mobile robots in the exploration problem (Batalin and Sukhatme, 2003a). The embedded nodes offer a suggested un-explored direction for the mobile robots to follow.

Mobile robots have also been used to support embedded networks. Lamarca et al. use mobile robots to continually calibrate a sensor network (LaMarca et al., 2002). Rahimi et al. present an approach for power harvesting in sensor networks by exploiting mobility (Rahimi et al., 2003). Corke et al. use a UAV to deploy and maintain the connectivity of a sensor network (Corke et al., 2004).

Acknowledgments

We would like thank our collaborators on the GNATs project, Victor Bigio, Eric Dodson, and Arya Irani. Also, we would like to acknowledge the National Science Foundation for funding under award #0326396.

Notes

1. **G**eorgia Tech **N**etwork/**N**ode(s) for **A**utonomous **T**asks

References

Batalin, M. and Sukhatme, G. (2003a). Coverage, exploration and deployment by a mobile robot and communication network. *Telecommunication Systems Journal, Special Issue on Wireless Sensor Networks*.

Batalin, M. and Sukhatme, G. (2003b). Sensor network-based multi-robot task allocation. *Proceedings of International Conference on Intelligent Robots and Systems (IROS 2003).*

Batalin, M., Sukhatme, G. S., and Hattig, M. (2004). Mobile robot navigation using a sensor network. *Proceedings of the IEEE International Conference on Robotics and Automation*, pages 636–642.

Corke, P. I., Hrabar, S. E., Peterson, R., Rus, D., Saripalli, S., and Sukhatme, G. S. (2004). Autonomous deployment and repair of a sensor network using an unmanned aerial vehicle. *Proceedings of the IEEE International Conference on Robotics and Automation*, pages 3602–3609.

Koenig, S., Szymanski, B., and Liu, Y. (2001). Efficient and inefficient ant coverage methods. *Annals of Mathematics and Artificial Intelligence*, 31:41–76.

LaMarca, A., Brunette, W., Koizumi, D., Lease, M., Sigurdsson, S. B., Sikorski, K., Fox, D., and Borriello, G. (2002). Making sensor networks practical with robots. In *International Conference on Pervasive Computing.*

Li, Q., DeRosa, M., and Rus, D. (2003). Distributed algorithms for guiding navigation across a sensor network. *The 2nd International Workshop on Information Processing in Sensor Networks.*

O'Hara, K. and Balch, T. (2004a). Distributed path planning for robots in dynamic environments using a pervasive embedded network. In *Proceedings of Third International Conference on Autonomous Agents and Multi-Agent Systems.*

O'Hara, K. and Balch, T. (2004b). Pervasive *Sensor-less* networks for cooperative multi-robot tasks. In *Proceedings of 7th International Symposium on Distributed Autonomous Robotic Systems.*

Parunak, H. V. D., Brueckner, S., and Sauter, J. (2002a). Synthetic pheromone mechanisms for coordination of unmanned vehicles. In *Proceedings of First International Conference on Autonomous Agents and Multi-Agent Systems*, pages 449–450.

Parunak, H. V. D., Purcell, M., and O'Connell, R. (2002b). Pheromones for autonomous coordination of swarming uavs. In *Proceedings of First AIAA Unmanned Aerospace Vehicles, Systems,Technologies, and Operations Conference.*

Payton, D., Daily, M., Estowski, R., Howard, M., and Lee, C. (2001). Pheromone Robotics. *Autonomous Robots*, 11:319–324.

Rahimi, M., Shah, H., Sukhatme, G., Heidemann, J., and Estrin, D. (2003). Energy harvesting in mobile sensor networks. In *Proceedings of the IEEE International Conference on Robotics and Automation*, page to appear, Taipai, Taiwan. IEEE.

Wagner, I. A., Lindenbaum, M., and Bruckstein, A. M. (1999). Distributed covering by antrobots using evaporating traces. *IEEE Transactions on Robotics and Automation*, 15(05):918–933.

ROLE BASED OPERATIONS

Brian Satterfield, Heeten Choxi, and Drew Housten
Lockhead Martin Advanced Technology Laboratories
3 Executive Campus, Cherry Hill, NJ 08002
bsatterf@atl.lmco.com

Abstract: The authors present an innovative approach to teaming humans and synthetic entities that leverages the concept of roles from research conducted in the business sciences. A teaming framework is presented that utilizes mission and team roles to allow for a natural integration of synthetic entities into existing human teams. Additionally, observations from experiments conducted within a testbed environment are described.

Keywords: Human-Robot Teaming, Roles, Autonomy, Military Application

1. Introduction

Military requirements and technology advancements are driving forces behind recent market trends that show an increasing usage of synthetic entities in the field. However, current state-of-the-art only allows warfighters to control either highly capable synthetic entities at the expense of their own effectiveness or to control much less capable synthetic entities while remaining effective. To maximize overall effectiveness, a paradigm change must occur which supports highly effective warfighters teaming with highly effective synthetic entities. We define teaming as the set of behaviors a group of entities perform while pursuing a common goal to coordinate knowledge regarding state, capabilities, needs, activities and obligations and to understand and reason about team structure, goals, tasks and the dependencies and roles within the team. We apply teaming as a solution to reduce warfighter overload, provide a force multiplier due to the addition of synthetic capabilities to a military team, and provide a means of integrating humans and synthetic entities that is natural to humans. There has been

L.E. Parker et al. (eds.),
Multi-Robot Systems. From Swarms to Intelligent Automata. Volume III, 283–289.

research in the teaming of robots and software agents with roles (Payne et al., 2000, Lewis, et al., 2003, Partsakoulakis et al., 2003), but much effort has been directed at multi-robot and multi-agent teaming with a smaller amount dedicated to human-robot interaction. We distinguish our research work from previous efforts in four ways: we enable teaming for humans, software agents and robotic systems; we have strived to adapt synthetic entities to humans and not vice versa; we believe synthetic entities must be realized as full team members and not "tools;" we utilize military requirements for teaming as a driving force for our architecture.

2. Research Question

Our research question was deceptively simple: How does one effectively team humans and synthetic entities in a military context? The simple answer was: by using techniques that humans have finely tuned over millennia.

Our research focus was driven by our intended application, military operations. Is there anything about military teams and their requirements that necessitate a change from "traditional" human teaming methods? Some differences are obvious. Individual members of military teams have an elevated personal risk, both physical and mental, and associated levels of stress. They require more intimate knowledge of teammate abilities, must adapt more quickly to internal and external events, and have a greater chance of operating outside of optimum configuration regarding skill matching to tasks due to personnel loss. We set out to understand how humans, who provide the best available example of teaming, behave in a team context and identify any available models of human teaming.

3. Approach

We wanted an approach grounded strongly in how humans currently perform teaming. Dr. Belbin's work on high performance management teams provides models of human teaming (Belbin, 1993). Dr. Belbin studied some of the top managers from around the world and their behavior patterns within a team context over a period of nine years. After collecting data, he noticed "clusters" of behavior, which he described in terms of roles that humans naturally gravitate toward within a team.

We collected over 25 U.S. Defense Department operational available scenarios involving unmanned vehicles and focused on missions related to cooperative reconnaissance. By analyzing these scenarios we noticed behavior "clusters" similar to the type Belbin discovered but unique to the military domain. To make the distinction, mission roles are concerned with mission execution; team roles are concerned with team management and maintaining team cohesion. Our complete set of roles pertaining to Cooperative Reconnaissance for mission and team are listed in Table 2.

Together, the team and mission roles define the entities in the team and their relationships regarding both team and mission management. The following is a set of features we have defined for roles thus far: (1) Explicit Coordination points, (2) Skills, (3) Responsibilities, (4) Team reporting structure, (5) Role dependencies, (6) Communication between roles.

4. Results

We have developed Role-Based Operations (RBO), a system that uses Roles to support teaming between humans and synthetic entities. RBO is a implemented system, similar to those covered in (Partsakoulakis et al., 2002, Tambe, 1977). The architecture for RBO augments current capabilities of synthetic entities to allow them to be "teamable" with humans. We implemented an initial set of roles to perform multiple missions depending on which roles are enabled. We conducted informal tests of our system to determine if roles could be used to effectively support heterogeneous teaming.

5. Research Results

Roles provide an abstraction between the type of entity—human, agent, or robot—and their responsibilities and capabilities (Partsakoulakis et al., 2003). The abstraction is not perfect, but it does allow humans to utilize natural methods of teaming. By adapting synthetic entities to use roles, they can contribute to teams and can use methods similar to humans to reason about team behavior. In this way, synthetic entities are required to adapt to humans in RBO.

Roles facilitate heterogeneous entity teaming by providing: (1) a description of an entity's capabilities, (2) a description of a team's expectations of the entity's behavior, (3) increased adaptation and responsiveness for the team, and (4) reusability for multiple missions.

Table 1. The roles we developed are based upon Dr. Belbin's research and numerous DoD scenarios that described Cooperative Reconnaissance missions.

	Role	**Description**
Team	Coordinator	Directs the action of team members
	Monitor	Observes the team and makes inferences
	Resource Investigator	Determines what is available and what can be done through communication within the organization
	Team Security	Protects and maintains the welfare of all team assets
	Maintenance	Performs routine tasks ensuring the proper functioning of assets.
Mission	Observer	Utilizes sensors to report or record
	Weapons	Executes a call for fire on a specific target
	Analyzer	Transforms data into information

Roles describe the capabilities of humans, robots and agents as well as the expectations a team has of an entity, e.g. an entity performing a team security role will have a set of expected behaviors. These expectations can be relied upon and used to predict behavior throughout the team regardless of current communications. Our role concept supports composition and inheritance. The Team Security role displayed in Figure 1 is composed of an Observer, Analyzer, and Weapons role, showing how more complex roles can be composed of more basic roles providing a new set of responsibilities without new development. An Observer role can be extended with child roles such as an Electro-optical Observer. The child role inherits the capabilities and responsibilities of their parent, but is more expressive.

In (Fong et al., 2004) a set of metrics to measure Human-Robot Interaction are provided. Measures of Robot Performance include Self-awareness, Human-awareness, and Autonomy. The use of roles contributes to self-awareness and human-awareness by making a synthetic entity aware of its responsibilities in a mission and the team's expectations of that entity. Our use of roles appeared promising on paper, and we created a prototype of RBO to test its effectiveness on real heterogeneous teams.

6. Experimental Results

A heterogeneous teaming testbed was created to test if RBO can effectively team humans and synthetic entities in a military context. The testbed consists of four iRobot Magellan Pro robots that serve as unmanned ground vehicles. Humans in our test use iPaqs equipped with a wireless card for communication.

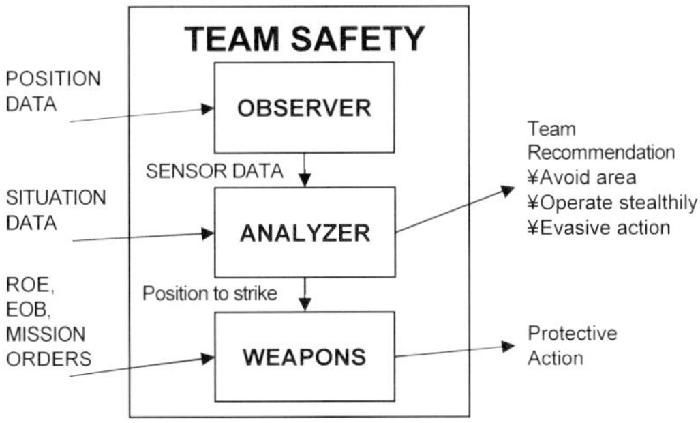

Figure 1. Roles can be composed of other roles to form more elaborate team and mission behaviors.

We created a simulated "Snatch and Grab" operation using our investigation of DoD military scenarios. We incorporated three roles in our scenario: a Seeker, Verifier, and Coordinator. Seekers are a composed role created by combining an Observer role with an Analyzer role. A Verifier is a specialization of an Analyzer role. The Coordinator role is the only team role used in our scenario; it is responsible for overseeing the operation including the mission start and stop directives to the team. In our testbed scenario both humans and robots are given the Seeker role, while the Verifier and Coordinator roles are given only to humans. In Figure 3, Robot Seekers and Human Seekers are both expected to search the area of interest. In order to insure that the Coordinator is getting correct information, the Verifier role is used to verify data the robot seeker sends out. If the robot correctly analyzed its data, the Verifier forwards its analysis of the data to the Coordinator. The Coordinator takes the data it gets and determines how to act on it.

Roles were a positive contribution to heterogeneous entity teaming in our experiments. Humans were able to understand their roles, the robots roles, and how the different roles were related to each other. Robots were not as efficient and effective as humans, but this was a problem due to the autonomous capabilities of the robots, and not the use of Roles. Also, the use of a Verifier role provided a simple way for a human to collaborate with a robot to increase the robot's effectiveness. The Verifier was able to filter out false id's the robot made, while the robot reduced the load on the Verifier by only sending data to the Verifier when it believed it found something, instead of every time it took a picture.

7. Future Work

Our research has shown that roles are an excellent way to capture information with respect to the team and its mission. However, we have identified many additional features that would increase team performance and two additional teaming concepts that we believe will provide a more complete solution for teaming in military operations.

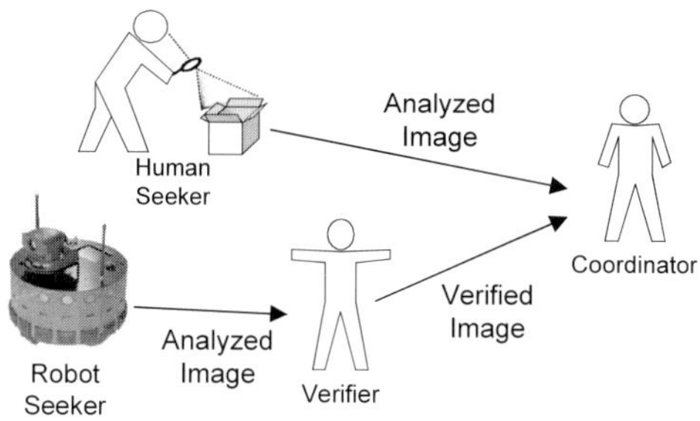

Figure 2. Synthetic entities perform roles suited to their sensing capabilities while humans play roles that require cognitive skills and high-level analysis.

Increasing the intelligence and reasoning ability of roles on synthetic entities will help increase self-awareness and team-awareness. Being able to map entities to roles using an effectiveness and efficiency rating for required capabilities should help facilitate planning with heterogeneous teams. Sharing world state information between heterogeneous entities is important. Currently we are working on the Zone Planning System that assigns semantic data to geographic areas to fill this need. Also, incorporating modes—such as attack mode, defense mode, and stealth mode—should be useful in improving the adaptability of synthetic entities while constraining their behavior to models human team members can understand.

Our future plans are to incorporate these new concepts into new scenarios, adding additional missions to our current selection. We also plan to conduct additional experiments that measure the performance of RBO. One aspect of our research that the current experiment does not measure is the effectiveness of roles in mission adaptation. Our current experiments also use informal methods to measure success; we plan to use the common metrics described in (Fong et al., 2004) for future experiments.

References

Belbin, M. (1993). *Team Roles at Work*, Butterworth-Heinemann.

Fong, T., Kaber, D., Lewis, M., Scholtz, J., Schultz, A., Steinfeld, A. (2004). *Common Metrics for Human-Robot Interaction*. Unpublished.

Lewis, M., Sycara, K., Payne, T. (2003). Agent roles in human teams. Proceedings of the AAMAS-03 Workshop on Humans and Multi-Agent Systems.

Partsakoulakis, I., Vouros. G. (2002). *Roles in Collaborative Activity*. In I. Vlahavas and C. Spyropoulos, editors, Methods and Applications of Artificial Intelligence, Second Hellenic Conference on AI, LNAI 2308, pages 449-460.

Partsakoulakis, I., Vouros, G. (2003). Roles in MAS: Managing the complexity of tasks and environments. Multi-Agent Systems: An Application Science.

Payne. T., Lenox, T., Hahn, S., Sycara, K., Lewis, M. (2000). Agent-based support for human/agent teams. Proceedings of the Software Demonstration; ACM Computer Human Interaction Conference; The Hague, Netherlands.

Tambe, M. (1997). *Towards Flexible Teamwork*. Journal of Artificial Intelligence Research, 7:83-124.

ERGODIC DYNAMICS BY DESIGN: A ROUTE TO PREDICTABLE MULTI-ROBOT SYSTEMS

Dylan A. Shell, Chris V. Jones, Maja J. Matarić
Computer Science Department, University of Southern California
Los Angeles, California 90089-0781, USA
dshell@cs.usc.edu, cvjones@cs.usc.edu, mataric@cs.usc.edu

Abstract

We define and discuss a class of multi-robot systems possessing ergodic dynamics and show that they are realizable on physical hardware and useful for a variety of tasks while being amenable to analysis. We describe robot controllers synthesized to possess these dynamics and also physics-based methodologies that allow macroscopic structures to be uncovered and exploited for task execution in systems with large numbers of robots.

Keywords: Multi-robot systems, Ergodicity, Formal methods.

1. Introduction

Multi-robot systems can both enhance and expand the capabilities of single robots, but robots must act in a coordinated manner. So far, examples of coordinated robot systems have comprised of largely domain-specific solutions, with few notable exceptions. In this paper we describe our ongoing work on the development of formal methodologies for synthesis of multi-robot systems that address these issues in a principled fashion.

We focus here on inter-robot dynamics, the roles played by those dynamics toward task achievement, and their implications in feasible formal methods for synthesis and analysis. We describe automated synthesis of controllers that capitalize on so-called *ergodic* dynamics, which enable mathematical arguments about system behavior to be simplified considerably. Sensor-based simulations and physical robot implementations show that these controllers to be feasible for real systems. We further suggest that this approach will scale to systems with large numbers of robots.

L.E. Parker et al. (eds.),
Multi-Robot Systems. From Swarms to Intelligent Automata. Volume III, 291–297.

2. Related formal methodologies

Formal methodologies for synthesis and analysis of multi-robot systems differ based on the type of systems they aim to address. One successful method for analysis of swarm systems is based on the theory of stochastic processes: for example, in the phenomenological modeling and analysis of multi-robot cooperative stick-pulling, a macroscopic difference equation for the rates of change of each type of robot state is derived from the stochastic master equation and sensor-based simulations are used to estimate parameter values (Martinoli et al., 2004). An extension to the same theory (but using continuous differential equations instead) allows adaptive systems to be modeled (Lerman and Galstyan, 2003). When applied to foraging, the analysis enabled system design improvements. Explicitly coordinated systems are typically addressed at the algorithmic level, such as in the Computation and Control Language (Klavins, 2003) and formal studies of multi-robot task allocation methodologies (Gerkey and Matarić, 2004). Also related is Donald's (1995) Information Invariants Theory, and Erdmann's (1989) studies of the advantages of probabilistic controllers.

3. Behavioral configuration space and ergodicity

The physical configuration space, common in robotics for representing physical arrangements, can be augmented to include additional dimensions for each of the robot's internal control variables that observable behavior. We call this the *behavioral configuration space* (BCS). It is a useful mental representation for a multi-robot system and for reasoning about the overall system dynamics. For practical applications, we will only consider particular subspaces, never the full configuration space.

The BCS of a single robot consists of dimensions for the physical configuration (e.g,. the pose variables, and velocities if necessary) and dimensions for internal state variables (continuous or discrete values within memory). The range of each dimension is determined by constraints on state variables. The BCS of an ensemble of robots is constructed from essentially a Cartesian product of individuals spaces and the spaces of movable obstacles, etc. The constraints (e.g., two robots simultaneously occupying the same location) subtract components from this product. Couplings between the robots via communication channels, mutual observation, etc., further restrict this space.

The global state of the multi-robot system at any specific time can be represented by a point in BCS and likewise, the time-evolution of the system, as a trajectory. A system that exhibits *ergodic dynamics* completely visits all parts of the configuration space with probability that is dependent only on the volume of that part of the space. Long term history is unimportant in predicting the dynamical behavior because the system "forgets" previous trajectories.

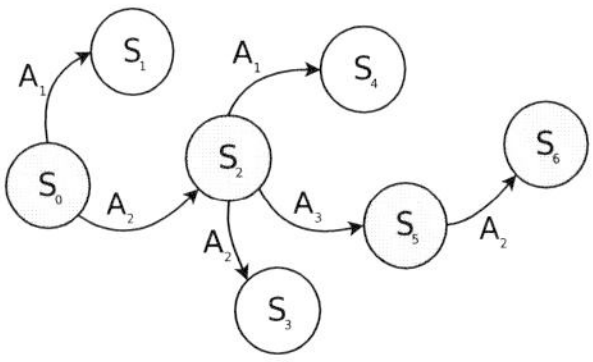

Figure 1 The Markovian *world* as states with action transitions. The sequential *task* specification is a particular sequence of world transitions. The task is shown here shaded within the world specification.

Time averages of some system property (over a duration longer than the underlying dynamics timescale), are equal to (configuration) spatial averages. Few useful robotics systems are entirely ergodic, but various sub-parts of the BCS may be ergodic. The next section describes one such system.

4. Automated synthesis for sequential tasks

Jones and Matarić (2004a, 2004b) have developed a framework for automatic and systematic synthesis of minimalist multi-robot controllers for sequential tasks. The framework consists of a suite of algorithms that take as input a formal specification of the environmental effects, the task requirements, and the capabilities of the robots. The algorithms produce either provably correct robot controllers, or point to the exact scenarios and task segments which make (algorithmically) guaranteed task completion impossible. The type of controller and prospect of successful task execution depend on the capabilities of the individual robots. Current options include the possibly of broadcast inter-robot communications (Jones and Matarić, 2004a), and a local memory on each of the robots permitting non-reactive controllers (Jones and Matarić, 2004b). Two complementary analysis techniques allow various statistical performance claims to be made without the cost of a full implementation and exhaustive experimentation.

We do not provide full details of the framework here, but instead focus on the (non-obvious) role of ergodic dynamics in the work. The framework uses a set of states S to denote the possible states that the (assumed to be Markovian) world can be in. The set A contains actions which act upon the world state, producing state transitions defined in some probabilistic manner (see Figure 1). A particular sequence of states, say $T, (T \subset S)$ makes up the task. In actuality the robots are only interested in producing the single task sequence, and thus only those transitions need to be stored. Thus, S is never stored or calculated, only T need be kept, and $|T| \ll |S|$. Robots then move around the environment making observations, perhaps consulting internal memory or listening to the broadcast communication channel if suitably equipped. If a robot has sufficient information to ensure that the performance of a particular action (from A) can only result a world transition that is part of the task (i.e., result in a state in T) then it may perform that action.

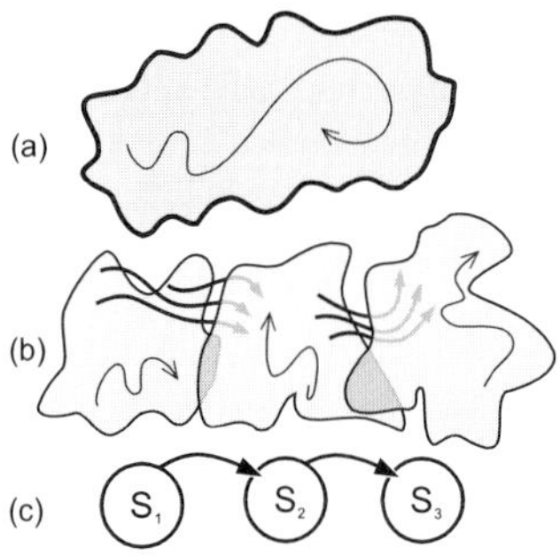

Figure 2 Three different views of the system dynamics: (a) A trajectory within the entire behavioral configuration space; (b) A hypothetical projection of the configuration space showing planar subspaces for behavior not part of S, and actions from A connecting these spaces; (c) The abstracted S representation as used by the described formalism.

Returning to the notion of behavioral configuration space, each of the world states in S represents entire subspaces of the overall system's space. Figure 2 shows that the entire behavioral configuration space as it fits into this formalism. A hypothetical projection of this entire (huge) configuration space separates the configurations into subspaces, each subspace representing a single state in T. Actions (from A) evolve the world state and hence transition the system from one subspace to another. *We design the system so that within each subspace the dynamics are ergodic.* Work that has used this formal framework ensured this property by having the robots perform randomized exploration policies. The randomized strategy needs to have sufficient effect to overpower other systematic biases in the system that could produce large scale effects and ignore some part of the configuration space.

Both controllers with memory (Jones and Matarić, 2004b) and ones endowed with communication capabilities (Jones and Matarić, 2004a) were demonstrated in simulation and on physical hardware in a multi-robot construction domain. The task involves the sequential placement of colored cubes into a planar arrangement. The sequence T contains simply the required evolution of the structure, actions A being the placement of an individual cube. Referring back to Figure 2; in the construction domain the motions within each subspace are random walks by the robots, and the transitions between spaces are cube placement actions.

Analytical techniques developed in order to predict task execution are aided by the ergodic components of the robots behavior in this domain. One example is in the macroscopic model (Jones and Matarić, 2004b, pp. 4–5) applied to this formal framework. This model calculates the probability of successful task completion by calculating a large multiplication of all possible memory states that get set, in each possible world state, after each possible observation, calculating the probability that only the correct action will result and includes terms for when actions may result in other, or null, world transitions. A fundamental assumption for that calculation is that no "structure" in the world results in the observation and action sequences that correlate. When endowed with navigational controllers that have ergodic dynamics, we know that this is true because the observation of an ergodic system at N arbitrary instants in time is statistically the same as N arbitrary points within the behavioral space (McQuarrie, 1976, pp. 554).

This section has demonstrated that dynamics with a high degree of ergodicity are achievable on physical robot systems. They can play a role in systems for which analytical methods exist, and as a very simple form of dynamics they can aid in simplifying particular aspects of system design.

5. Large-scale multi-robot systems

We consider large-scale multi-robot systems those with robots on the order of thousands. In spite of the fact that manufacturing and tractable simulation remain open challenges, a variety of tasks have been proposed for systems of this type. Increasing the number of robots increases the total number of degrees-of-freedom in a system, and results in a highly dimensional BCS. Coordination approaches that couple robot interactions as loosely as possible are most likely to scale to large sizes.

Mathematical techniques employed in statistical mechanics are useful for establishing the relationship between microscopic behavior and macroscopic structures (McQuarrie, 1976). Typical system sizes for classical work are significantly larger ($\sim 10^{23}$) than the numbers currently conceivable for robots. In the case of large (or infinite) systems, interesting macroscopic structures can result even from ergodic local dynamics. global structures like equilibrium phases, phase transitions, coexistence lines, and critical points are widely studied in thermodynamics. Recent work attempts to reformulate many of these classical notions for systems with fewer entities (Gross, 2001).

We are pursuing a methodology for coordination of large-scale systems through the study of a small set of mechanisms for producing general macroscopic phenomena. One candidate mechanism is a protocol for achieving consensus. The Potts (1952) model is illustrative; it is an archetypal magnetic spin system that models interactions between particles at a number of fixed locations within a graph or lattice. The Ising model (a specific Potts model) has also been used to model gas flow. Neither model is a perfect fit for robots, but illustrates macroscopic structure from simple interactions.

Mapping the spin interactions at spin sites to robots allows for the development of a communication algorithm that possesses ergodic dynamics (and an energy conservation constraint) that permits the definition of a partition function $\mathcal{Z}$ that can be solved using a numerical method for pseudo-dynamics simulation (or in trivial cases analytically). This admits a prediction of global behavior because exhaustive parameter variations enable construction of a phase diagram. In the case of the Potts and Ising models this phase diagram is well known. Particular regions of the phase space in the Ising model represent regions of maximal order. For robots this means unanimity; consensus is reached through a second-order phase transition.

The ability to prescribe ergodic dynamics for large-scale robot systems makes those analytical approaches that focus only on constraint space topology feasible for predictions of global structure. This means that task directed actions can be tackled directly from a macroscopic perspective.

6. Summary and Conclusion

We have taken a dynamics-centric approach to describing multi-robot behavior. This view has suggested that the notion of ergodicity may be useful within a robotics context, something that we have demonstrated throughout the paper. After defining a behavioral configuration space, we demonstrated that subspaces in which the robot dynamics are essentially ergodic can be used to produce meaningful behavior, and allow automated synthesis techniques to focus on a small set of task-oriented states, rather then the entire ensemble configuration space. Also, in at least one case, ergodicity simplifies analysis of system behavior. Implementations on physical and simulated robots show that ergodicity is indeed achievable in the real world. Future promise of this general approach is suggested in a discussion of large-scale multi-robot systems.

Acknowledgments

This research was conducted at the Interaction Lab, part of the Robotics Research Lab at USC and of the Center for Robotics and Embedded Systems. It was supported by the Office of Naval Research MURI Grant N00014-01-1-0890.

References

Donald, B. R. (1995). On information invariants in robotics. *AI*, 72(1-2):217–304.

Erdmann, M. A. (1989). *On Probabilistic Strategies for Robot Tasks*. PhD thesis, M.I.T.

Gerkey, B. P. and Matarić, M. J. (2004). A formal analysis and taxonomy of task allocation in multi-robot systems. *International Journal of Robotics Research*, 23(9):939–954.

Gross, D. H. E. (2001). *Microcanonical Thermodynamics: Phase Transitions in "Small" Systems*. World Scientific Lecture Notes in Physics - Volume 66. World Scientific, Singapore.

Jones, C. V. and Matarić, M. J. (2004a). Automatic Synthesis of Communication-Based Coordinated Multi-Robot Systems. In *Proc. of the IEEE/RSJ International Conference on Intelligent Robots and Systems (IROS-04)*, Sendai, Japan.

Jones, C. V. and Matarić, M. J. (2004b). Synthesis and Analysis of Non-Reactive Controllers for Multi-Robot Sequential Task Domains. In *Proc. of the International Symposium on Experimental Robotics*, Singapore.

Klavins, E. (2003). Communication Complexity of Multi-Robot Systems. In Boissonnat, J.-D., Burdick, J., Goldberg, K., and Hutchinson, S., editors, *Algorithmic Foundations of Robotics V*, volume 7 of *Springer Tracts in Advanced Robotics*, pages 275–292. Springer.

Lerman, K. and Galstyan, A. (2003). Macroscopic Analysis of Adaptive Task Allocation in Robots. In *Proc. of the IEEE/RSJ International Conference on Intelligent Robots and Systems (IROS-03)*, pages 1951–1956, Las Vegas, NV., USA.

Martinoli, A., Easton, K., and Agassounon, W. (2004). Modeling Swarm Robotic Systems: A Case Study in Collaborative Distributed Manipulation. *IJRR*, 23(4):415–436.

McQuarrie, D. A. (1976). *Statistical Mechanics*. Harper and Row. reprinted by University Science Books, Sausalito, CA., USA in 2000.

Potts, R. B. (1952). Some generalized order-disorder transformations. In *Proc. of the Cambridge Philosophical Society*, volume 48, pages 106–109.

Author Index